AF680775

FNNR
Foundation for Neurofeedback
& Neuromodulation Research

FNNR Publications

Sokhadze, E., Morosoli, T., Frederick, J., Evans, J. (2023). *International Trends in Neurofeedback Volume 1.* Foundation for Neurofeedback and Neuromodulation Research.

Longo, R. E., & Bingham, B. (2022). *Beginning Neurofeedback in Your Practice: From Intake to Discharge.* Foundation for Neurofeedback and Neuromodulation Research.

Soutar, R., & Longo, R. E. (2022). Doing Neurofeedback: An Introduction (2nd Ed.). Foundation for Neurofeedback and Neuromodulation Research

Bester, Helena (2020). Neurofeedback: The Non-invasive Alternative. Foundation for Neurofeedback and Neuromodulation Research.

Longo, R. E., & Soutar, R. (2020). *Becoming Certified in Neurofeedback: A Guide to the Neurofeedback Mentoring Process for Mentors and Mentees.* Foundation for Neurofeedback and Neuromodulation Research.

Sokhadze, E., & Casanova, M. F. (Eds.). (2019). *Autism Spectrum Disorder: Neuromodulation, Neurofeedback, and Sensory Integration Approaches to Research and Treatment.* Foundation for Neurofeedback and Neuromodulation Research.

Martins-Mourao, A., & Kerson, C. (Eds.). (2017). *Alpha-Theta Neurofeedback in the 21st Century: A Handbook for Clinicians and Researchers* (2nd ed.). Foundation for Neurofeedback and Neuromodulation Research.

Soutar, R., & Longo, R. E. (2011). *Doing Neurofeedback: An Introduction.* The ISNR Research Foundation.

Hammond, D. C., & Gunkelman, J. (2011). *The Art of Artifacting.* The ISNR Research Foundation.

Carmichael, J. (2011). *Multi-Component Treatment for Post-Traumatic Stress Disorder, Including Strategies From Clinical Psycho-Physiology and Applied Neuroscience.* The ISNR Research Foundation.

Donaldson, S. (2012). *The Other Side of the Desk.* The ISNR Research Foundation.

Managing and Integrating Trauma With Alpha/Theta Neurofeedback

Richard Soutar

Publisher:

The FNNR

The Foundation for Neurofeedback & Neuromodulation Research

101 Stewart Street Hendersonville, NC 28792

http://www.theFNNR.org

Correspondence: admin@theFNNR.org

Copyright © 2025 - Richard Soutar

All rights are reserved under international and Pan-American copyright conventions. No part of his publication may be reproduced, stored in a retrieval system, or transmitted in any form or by any means, electronic, mechanical, photocopying, recording, scanning, or otherwise, except as permitted under Section I07 or I08 of the I976 United States Copyright Act, without either the prior express written permission of the author or publisher or authorization through payment of the appropriate per-copy fee, except for brief quotations in critical reviews.

Limit of liability/Disclaimer of Warranty: The content, guidance, recommendations, and practice guidelines outlined in this book may not reflect the views and/or opinions of The Foundation for Neurofeedback and Research (FNNR). While the publisher and author have used their best efforts in preparing this book, FNNR makes no representations or warranties with respect to the accuracy or completeness of the contents of this book and specifically disclaims any implied warranties of merchantability or fitness for a particular purpose. No warranty shall be created or extended to FNNR by sales representatives or written sales materials. The advice and strategies contained herein may not be suitable for your situation. You should consult with a professional where appropriate. Neither the publisher nor author nor ISNR shall be liable for any loss of profit or any other damages, including but not limited to special, incidental, consequential, or other damages.

All marketing information is presented with the assumption that the clinician is applying suggestions appropriate to their healthcare license, HIPAA regulations and information shared by or about clients is with their permission.

This book is not meant to substitute for medical or psychotherapeutic advice or treatment

Publisher: The FNNR ,The Foundation for Neurofeedback & Neuromodulation Research

101 Stewart Street Hendersonville, NC 28792

http://www.theFNNR.org. Correspondence: Admin@theFNNR.org

Edited by Jon Frederick

Layout Design: Shea Connor. Correspondence: publishing@theFNNR.org

Cover by: Shea Connor
Distributed by: BMED Press, LLC
Managing and Integrating Trauma With Alpha/Theta Neurofeedback

ISBN: 979-8-9889374-3-2

Table of Contents

Introduction

Around the 1980s a new conceptualization of trauma began to emerge in psychiatry and the behavioral sciences. Research projects like the Adverse Child Experience Study (ACEs) has led to the general acknowledgement of trauma as a key factor in the manifestation of both physical and mental health. This is a game changer of tremendous importance. Trauma arises from the interplay of social and psychological forces that contribute to the emergence of symptoms and behaviors that in turn feed back into the social system and shape cultural norms and mores. Since the late twentieth century we have learned that trauma is intergenerational, and is handed down to succeeding generations through mechanisms of transgenerational behavior within the family system and the socialization process. We have learned how trauma shapes biology and vice versa in a bi-directional dance of incredible complexity, and much of the dance is involuntary and unacknowledged.

Trauma is an inescapable part of life and an eternal component of the human condition. It is the dark and fearful beast we must all face in the process of socialization and the formation of our identity. War, poverty, and social injustice feed off of it and in turn the behaviors arising from trauma ensures their continuation. I believe it is the fundamental challenge we must face as a species, if we are going survive. How we respond to trauma and the social and political forces it unleashes will inevitably determine the level of threat we are to ourselves and to life on the planet during this sixth extinction.

As functional MRI (fMRI) and quantitative EEG (qEEG) provide us with a new means of identifying the nature of the mechanisms of trauma in terms of human physiology and the central nervous system, they also suggest solutions. One such solution is the altering of the distortions of structure and function of the brain that trauma produces, and normalizing them through conscious feedback utilizing these same technologies, known as neurofeedback. One such type of neurofeedback, known as EEG biofeedback, is emerging from its quiet clinical gestation over the past 40 years as one of the most viable of these technologies. I have found that literature on this topic has grown immensely at the international level in the last 10 years and so has the enthusiasm.

Within the field of neurofeedback one technique was uncovered at the outset that used feedback to guide individuals into a twilight state that provided the an unusual opportunity to integrate trauma very effectively. It developed under the rubric of alpha/theta training, as these were the two designated names of the brain wave frequencies employed in the process. This technique has stood the test of time as a powerful clinical tool for integrating trauma, especially when used in conjunction with other clinical skills and technologies, and it is becoming widely used in clinics worldwide. It is exceedingly difficult to apply rigorous controlled group designs to clinical methods because of the multiplicity of variables and individual differences of subjects. There is, however, some good research pointing to its potential efficacy but its mechanism of action is only beginning to be understood. On the qualitative side, clinicians who encounter the technology are almost universally stunned by its apparent efficacy.

As a consequence of the foregoing, my purpose in writing this book is to shine a light on the power of this method of neurofeedback. It is to identify and outline the nature of psychophysiological trauma according to the recent literature, in addition to outlining and identifying a powerful solution in the form of neurofeedback. It is also an effort to examine alpha/theta neurofeedback in as full detail as possible given what we know now based on the present science.

Chapter 1

The Tangled History of Understanding Trauma and Its Treatment

Early History

The first and most salient recognition in modern times of the importance and impact of psychological trauma on the individual and their daily functioning was probably the publication of the now renowned seminal work on combat trauma "The Traumatic Neurosis of War" by Abram Kardiner published in 1941. However, the recognition of war based psychological trauma goes back to ancient times, being mentioned in the Epic of Gilgamesh as well as by the Greek historian Herodotus and in the reports of army physicians of the French Revolutionary wars (1792–1800), just to name a few sources (Crocq & Crocq, 2000). The Industrial Revolution also gave rise to some unique civilian man-made disasters, such as railway accidents, where the psychological symptoms and behavior of survivors puzzled physicians. The term "traumatic neurosis" began to emerge along with attending arguments regarding whether the symptoms were a result of organic or emotional causes.

By World War I, the medical community and the military command were forced to confront the bewildering issue that they had more psychological casualties than physical ones due to the constant shelling that soldiers had to endure. Typical symptoms included sudden muteness, deafness, general tremor, constant nightmares, inability to stand or walk, episodes of loss of consciousness and convulsions. These phenomena challenged the dominant psychoanalytic theory of the time which argued that this type of hysteria could only come from childhood issues and the entrenched belief that events of severe emotional intensity could not result in such symptoms with apparently healthy adults. The loss of memory, smell and taste was also observed. Of particular interest in all these cases was the importance of the impact of the observed death and mutilation of comrades, as Herodotus had mentioned. The British coined the term "shell shocked" to manage the issue in the media and the treatment that evolved was called the "forward treatment."

Forward treatment involved treating the soldiers near the front lines with the support of their comrades and officers with the expressed expectation of returning them to battle. This approach emerged when it was discovered that evacuated soldiers had a very poor prognosis compared with those treated in centers near the front lines. There was also a general concern regarding the malingering of those soldiers who might take advantage of the diagnosis to get discharged with full disability pensions and the possibility that large numbers might desert through this diagnosis. The British military was so concerned that they actually went to the extreme of executing 346 soldiers who presented with this symptomatology in order to send a message to the troops regarding their concerns over this issue. Oddly, the more successful methods of treatment that emerged during WWI were quickly forgotten.

In this historical record we can already begin to see the social and political threads of society struggling to weave an acceptable narrative of this unexpected and confusing phenomenon. There was little room in the popular concept of the "warrior" for this type of response to war. The narrative, however, would not acquire the input of a maturing psychological understanding of human behavior until the reluctant declaration in 1980s DSM of the diagnosis of PTSD. The issue did not receive its fully deserved recognition until finally in the 21st century with the recognition of complex trauma by the WHO. So, the historical record tells our acceptance of the possibility of enduring psychological trauma due to social and industrial tragedies met with great resistance is relatively recent. This was clearly in part due to the ongoing threat of acknowledging the full psychological consequences with respect to war. There was grave concern regarding the potential recognition of the failure of national moral fiber and how that might inevitably lead to losing a war—a social amplification of panic in the ranks that every general in history feared. The reluctance of industry was clearly due to greed and the cost of improving work conditions and the increased cost of workers compensation.

During WWII, this forgotten information regarding the phenomenon and its treatment resulted initially in only 5% of troops experiencing these symptoms returning to battle. It wasn't until 1943 that the Forward Treatment was reinstituted resulting in 50%–70% of troops being returned to duty. The term "exhaustion" was deployed along with "operational fatigue" to reduce the insinuation of anything more severe for purposes of national and military morale. At the time, Kardiner noted while conducting research in a Bronx Veterans hospital, that many veterans who were patients "were overtaken by a sense of futility; they became withdrawn and detached"—traumatic neurosis developed resulting in a chronic vigilance and sensitivity to threat. He recognized and acknowledged that there were somatic as well as psychological aspects of the phenomenon, which echoes the concept of "the body keeps the score" promoted by van der Kolk (2014) in recent times. Grinker and Spiegel (1945) also examined this phenomenon at the time but they developed a barbiturate-based treatment in the hopes of providing a rapid and simple solution to the problem. This opened up thinking regarding the development of the drug-based solution which dominates psychiatry today.

Although the lessons learned from WWII were applied in the forward area of the Vietnam War, alcoholism and drug abuse were rampant among the troops and almost half those sent into action required psychological help for delayed and chronic PTSD after the war. The full nature of PTSD was still poorly understood at the time but at least recognized and distinguished from "acute stress disorder," which reportedly most recovered from fairly quickly. Although many of the troops discontinued the use of drugs upon returning home, a significant number failed to effectively integrate back into society, becoming a burden on the resources of the time.

The picture painted by Crocq & Crocq (2000) is an invaluable snapshot of human history that lends depth to our understanding. The social and political denial is clearly a component and key feature which cannot be ignored, especially the century long avoidance of public debate throughout the 20th century despite the critical role it played in military operations. It likely plays a role in the nature of the manifestation of symptom typology as well as the treatment protocols. Was the incidence rate of post-Vietnam PTSD a different demographic than post WWII cultural denial? How does this play into the national reluctance to acknowledge complex trauma?

PTSD and Complex Trauma

The term PTSD and the concept came into existence in 1980 when a group of Vietnam vets, in concert with Robert Lifton, lobbied the

American Psychiatric Association to accept a cluster of symptoms common in vets. The VA was very slow to accept the concept according to van der Kolk who was deeply involved in research on trauma at the time (van der Kolk, 2014). Again, cultural and political forces conspired against him and others who recognized the high incidence rate among the general non-military citizenry. Acceptance of such a diagnostic category heralded monetary consequences for insurance providers as well as organizational restructuring for the veterans hospitals. The paradigm shift would also affect academia and research grants.

PTSD first appeared as an operational diagnosis in the DSM-III (1980) and was modified in the DSM-III-R (1987) and DSM-IV (1994). Despite efforts to expand the definition of trauma by van der Kolk and others, complex trauma was until very recently still subsumed under the rubric of PTSD in the DSM-V. Acknowledgement of a much higher incidence rate of emotional trauma in the public could have serious and perhaps undesirable implications for the insurance companies and service providers in general. It can profoundly affect their bottom line. Policy makers are typically intimidated by such sudden shifts in the DSM which can have profound institutional ramifications. There are political, professional, and financial concerns at stake.

In parallel with these developments, but not mentioned in the trauma literature, was the development of sophisticated implicit trauma models related to addictions and family systems. Codependency, in particular, was a topic of considerable interest in the 1980s which explored the traumatic effects of dealing with a family member who had an addiction, and the long-term intergenerational impact it had on development in the family system (Soutar, 1993). Addiction soon came to be broadly defined in these emerging models and included gambling, food, and sexual addiction. The impact of addictive behavior was documented in detail and the prevailing literature explored changes in family structure, family scripts, roles, and behaviors as well as communication styles and content. Early on investigators looked at the impact of prior generation's addictive behavior using genograms and began defining a transgenerational process focused on emotional dysfunction and trauma. Although the central role of trauma as a unitary underlying dimension was not as explicitly stated as at present, many authors in the 1980s and early 90s such as John Bradshaw (1988) outlined in detail the dimensions of transgenerational trauma and how it was perpetuated.

These authors and investigators were very familiar with the similarity of behaviors of veterans returning from war with self-medicating addictions and the non-military cases of emotional trauma. Their impact on their family systems and children's behaviors displayed the same traits. These authors also explored and revealed to the public the extensive traumatic childhood history of the majority of those in the criminal population. The impact of socio-economic status and race was also well documented in sociological studies with respect to dysfunctional and maladaptive behaviors emerging from traumatized or broken family systems (Olson, 1989). Alcoholics clearly emerged primarily from rigidly disengaged family systems (Soutar, 1996). What was still not so clearly understood was the role of physiology and neurology in the process and the impact it had on behavior.

More than half of people who seek psychiatric care have been assaulted, abandoned, neglected, raped, or exposed to family violence (van der Kolk, 2014). Family violence research has also focused attention on socio-economic status, addiction, war and other associated factors as key contributing factors in the emergence of violence in the family system. Today we know that one in three women and one in 10 men, 18 years of age and older, have experience domestic violence (Huecker et al., 2022). Sexual abuse studies also began in the 80s to emerge with recognition of the impact it had on the family system. As van der Kolk noted, the official literature in the sixties recognized an incidence rate on incest as approximately one in a million. Today we recognize the rate as far higher and even by the 1980s the National Institute of Justice reported a 10% incidence rate (1984) but it is recognized that this phenomenon is vastly underreported. Since we also now know today that experiences of sexual abuse are far more common than previously understood, these figures project a fairly widespread incidence of potential emotional trauma. Combined with reported incidence rates of physical abuse and emotional abuse, and based on these figures, it is likely we should consider complex trauma as fairly ubiquitous and represent it as a spectrum ranging from mild to severe.

The Physical Component of Trauma—The Body Keeps the Score

There was also considerable investigative initial work in the field of psychology on the developmental and physiological consequences of stress, which is a key component of trauma, in the 70s and 80s. Harry

Harlow (1958) did research with rhesus monkeys and found that given a choice between a wire surrogate mother that dispensed milk and a surrogate cloth mother, baby monkeys overwhelmingly preferred the cloth mothers. That a species so similar to us would prefer comfort over sustenance came as a shock to most theorists. This finding also demonstrated the importance of different aspects of caregiving with respect to cognitive and emotional development. The importance of emotional support with respect to the human developmental process was to be later emphasized in the findings of the renowned ACE study conducted between 1995 and 1997 (Finkelhor et al., 2015). This landmark study found both profound psychological and physiological consequences accrue from lack of emotional support from caregivers during childhood development.

Harlow also experimented with raising the baby monkeys in isolation and found that they matured into very cognitively underdeveloped and intensely emotionally disturbed monkeys. The monkeys who only had access to the wire mothers also developed serious digestive disorders. These research findings were predictive of the more recent findings regarding serious microbiome disturbances in pregnancy of mothers with ACE history (Hantsoo et al., 2019) as well as abandoned and abused children (Callaghan et al., 2020). The research by Harlow and others contributed significantly to our understanding of the importance of emotional nurturance and physical contact and significantly informed John Bowlby's (1982) attachment theory.

John Bowlby formulated the basic tenets of his theory regarding childhood development and trauma by drawing on concepts from ethology, cybernetics, information processing, developmental psychology, and psychoanalysis. Ainsworth and Bowlby (1991) together revolutionized our thinking about how separation, deprivation, and bereavement disrupt development and actually alter the central nervous system. This work in turn lead to the important work of Alan Schore in his book Affect Regulation and The Origin of the Self (Schore, 1994) which explored the importance of corticolimbic development in infants and early childhood with respect to coping with the stressful dynamics of human relationships as the individual matures. Along with neuroimaging research and the work of Teicher et al. (2016), who documented EEG and MRI abnormalities in traumatized children, this line of investigation has revolutionized our understanding of human behavior. This research indicates volumetric changes in the brain including aspects of structure, function and connectivity are profoundly impacted by emotional trauma. A recent large-scale study, N = 9270, replicates and further supports these findings (Jeong et al., 2021). It is clear that we grow and develop emotionally dysregulated brains under excessively stressful conditions. Unfortunately, most of the world outside the research community is still catching up. The majority of the public is unaware of these studies or implications and tends to minimize the impact of child abuse by reframing the definition, such as is found in many religious cults. Political concerns surrounding abortion rights also typically fail to take into account this aspect of parenting seriously. The public appears to assume that anyone in general who can hold down a job, can successfully raise a child. In fact, in our own clinic, when we attempted to refund parents $50 a week if they would attend a parenting course over the internet, we could still not enroll the majority of parents. Child abuse and dysfunctional family systems contribute as much, if not more, to the incidence rate of significant emotional trauma within the population, and yet our political will and economic support surrounding these issues fails to reflect the critical role they play with regard to social stability.

The work of Hans Selye on stress and the general adaptation syndrome, and Martin Seligman on learned helplessness, around the same time Bowlby and Ainsworth were investigating their theories, demonstrated the impact of stress and double binds on psychological behavior (Maier & Seligman, 2016). The studies by Herbert Benson at Harvard on the relaxation response (2000) also contributed significantly to our physiological understanding of the consequences of stress. These investigators demonstrated that stress can significantly impact any organism adversely regardless of age or developmental phase. The value of the insights arising from this paradigm is that reductions in stress can reduce and even reverse the psychophysiological patterns that undermine organic function.

Biofeedback in conjunction with lifestyle changes has the potential to provide individuals with the ability to control and mitigate the impact of stress on the biological system. Biometrics is flourishing today but it is mostly used to alter lifestyle and enhance performance and outside the clinic is used primarily by the self-disciplined professional class. The average citizen is unlikely to have the software skills, education, free time and monetary resources to take advantage of these technologies. However, simple practices such as breathing techniques and mindfulness can provide similar results but still require significant self-discipline and delayed gratification skills to mobilize. In a consumer society that emphasizes immediate gratification, the utilization of biofeedback remains problematic for the public.

TBI and Emotional Trauma

The impact of TBI in the form of multiple sub-concussive events on cognitive function and behavior is becoming more relevant to the development of emotional trauma as well. The recognition of a greater than previously acknowledged incidence rate of TBI, based on more sophisticated definitions grounded in neuroimaging, also revealed this as another source of trauma. TBIs apparently involve cognitive and mood problems that can have a significant potential impact on psycho-social behavior in general and consequently relationships and family systems (Montenigro et al., 2017; Sariaslan et al., 2016). Family violence and physical child abuse can often result in neurological disorder. The realization that multiple smaller concussive events, including frequently slapping a child, can result in neurological outcomes similar to one major concussive event has changed the picture of abuse. It has reframed the concept considerably. Either emotional abuse or physical abuse can also lead to self-destructive behavior in children such as head banging wherein children compulsively bang their heads against the wall or floor when distraught. Automobile accidents can lead to TBIs that reduce frustration tolerance levels, increase impulsivity and generate anxiety and depression in parental behavior patterns. This can lead to unintentional verbal and physical abuse by parents of their children. Industrial accidents, addiction, poverty and a host of other similar sociological events can result in neurological damage and stressors that lead to complex trauma and degrade human interactional exchanges. Unfortunately, these components of human behavior are for the most part underacknowledged or completely missing from public awareness and dialogue.

Anxiety and Depression as Key Measures of Trauma

It has become apparent, or it should have at this point in time, that much of what we call mental disorder has its source in emotional and developmental trauma now being presented under the rubrics of complex trauma and toxic stress. I would argue, based on 20 years of clinical experience, data collection and analysis, that a key measurable psychophysiological feature of socio-emotional and physical trauma, that can be easily and empirically obtained through qEEG and psychometrics, is severe anxiety and depression (Soutar, 2017a). It is rare to encounter a physical or emotional trauma that does not at least temporarily carry with it one of these fundamental dispositions of the human CNS. I would also argue that they are both psychological and physiological in nature and should not be considered mental disorders but rather natural responses of the human organism to stress. They are also excellent proxy measures of improvement and recovery from trauma when combined with sleep quality measures. Directly addressing the psychophysiological presentation of these two aspects of response to trauma through biofeedback in general and neurofeedback in particular has proven to be a powerful clinical antidote to the human suffering arising from complex trauma. At New Mind alone, we have over a decade of data from over 500 clinics demonstrating the validity and reliability of this technology when properly delivered through well-trained professionals.

As a consequence of the foregoing, I will present a clinical perspective as well as clinical practices that grew equally out of the above discussed literature as well as the more recent literature as these methods have important intrinsic value for therapists and provide a more developed clinical toolbox. The model here is inter-disciplinary and reflects a wide range of research and effective methods that grew out of that research. Prior to this discussion, however, I believe it is important to review what has not worked with respect to trauma and in particular the use of medications.

The Failure of Medications in the Treatment of Trauma

In 1987, the FDA approved Prozac, especially for depression, and over the next two decades the number of disabled mentally ill adults dramatically increased so that by 2007 it was more than double the rate in 1987 (Whitaker, 2010). In the same time period, the number of mentally ill children rose 35-fold so that it became the leading cause of disability in children. Drugs are critical for helping manage acute mental health problems, but they are a poor long-term solution and were never meant to be "the solution."

French scientists in the 1950s discovered chlorpromazine (Thorazine) made patients more tranquil, i.e., less agitated and delusional. The American Journal of Psychiatry published a study in 1968 demonstrating schizophrenics who receive the new drugs did better than those who received talk therapy (van der Kolk, 2014). This opened the door, according to van der Kolk, for psychiatrists to become "real scientists" with labs and empirical experiments. In fact, van der Kolk himself was, according to his own report, one of first psychiatrists to use lithium and was deeply impressed with its effects. Antipsychotics revolutionized psychiatry and reduced mental hospital occupancy from over 500,000 in 1955 to less than 100,000 by 1996.

In 1969 a WHO 5-year study that tracked schizophrenia outcomes in nine countries showed that the developing country outcomes that did not employ the use of antipsychotics has better outcomes than in the US. Follow-up studies confirmed the findings. Whitaker notes that by the 1980s research had shown that although the antipsychotics were effective, they created changes in the brain that unfortunately forced patients to remain permanently on the drugs. A further discovery was also made that over time patients were inevitably subject to tardive dyskinesia and eventually tardive psychosis. Neuroleptics, it turns out, were a trap with a built-in psychosis-inducing agent. By 1994 researchers were talking about tardive dementia and three-fourths of medicated schizophrenic patients 70 years and older were suffering a cognitive decline similar to Alzheimer's. By 1998 MRI findings were fully supporting this dire conclusion. In 2007 a study by Martin Harrow showed clearly over 15 years that those on antipsychotics did dramatically worse.

Benzodiazepines have a similar story. Between 1968 and 1981 valium ("mother's little helper") was the bestselling drug in the western world. Yet as it came into wide usage for anxiety, mental health hospitals and outpatient clinic admissions soared along with psychiatric emergency room visits. Other benzos appeared and many studies did not support efficacy and deployment soon revealed addictive features that were unanticipated, so that by 1975 they were classified as schedule IV drugs that were addictive. By the 1970s four times more people were addicted to benzos than heroin. Furthermore, the sleep-inducing effects of the drug did not last more than 2 weeks and anxiety-reducing effects didn't last beyond 4 months. Despite that, by 2002 they were still the primary psychiatric solution to anxiety and continue to be increasingly prescribed. Like antipsychotics, by 1983 the WHO reported that long term use increased anxiety, depression, cognitive impairment and resulted in general poor life quality.

Research in tricyclics by the 70s showed they were only slightly better than placebo. By 2008 a study in the UK showed that Prozac, Effexor, Serzone, and Paxil were barely more effective than placebo and not clinically significant in their effect. This was after a study years earlier at NIMH in 1984 found that 71% of depressed patients relapsed within 18 months of discontinuing medication. Studies at the time were showing long term effects of cognitive therapy were better and it was during this period that cognitive behavioral therapy (CBT) began to take shape. An additional study in 2006 revealed that most people with depression who are not medicated recover within one year. Today, depression treated with medication is the leading cause of disability according to NIMH.

DSM III brought the first diagnostic criteria for attention deficit disorder (ADD). Van der Kolk observed that it was ironic that in the forward of the book the authors of the DSM III acknowledged that it was so imprecise that "it should never be used for forensic or insurance purposes" (van der Kolk, 2014, p. 28). Then in 1987, ADHD was recognized in the DSM III-R. Ciba-Geigy, who owned and manufactured Ritalin, helped fund CHADD or Children and Adults with Attention Deficit Hyperactivity Disorder which has become the leading advocacy group. This was a marketing opportunity disguised as support for a well-intentioned group and a clear conflict of interest. Nevertheless, CHADD flourished and recognized stimulant medication and cognitive therapy as the primary evidence-based treatment for ADHD despite the fact that it fails to meet these standards according to the definitive study on the matter, the MTA cooperative study funded by NIMH. Russel Barkley, a putative expert on ADHD and a primary critic or NFB, fails to hold stimulant medication to the same evidentiary standard as he applies to NFB. Interestingly, he has received continual funding from the pharmaceutical industry (Pigott et al., 2021). What the research indicates is that what it does do is temporarily make children less disruptive.

A specific etiology and cause for this supposed disorder has never been found. Most diagnoses result from teacher complaints. The drug Ritalin is a dopamine reuptake inhibitor like cocaine except it clears more slowly. Dopamine receptors diminish during extended use of the drug. Serotonin and norepinephrine pathways are also affected. It has a seriously negative effects on emotional and social behavior but reduces excess movement—good for the teacher but not for the kid. Further research has shown it does not help academically in the long term and often results in progressive behavioral and social problems (Graf et al., 2013). In fact, the research indicates it doesn't help children learn at all (Pelham et al., 2022).

As Whitaker and van der Kolk note, the DSM redefined modern psychiatry and put the medical model firmly into place, increasing status and salaries of psychiatrists everywhere. From that time forward conflicts of interest have blossomed, and money flowed between the APA and the drug companies. The story has been retold in many publications in depth and the 20th century was a time when special interest profoundly influenced the diagnostic categories and defined the legitimate course of treatment while invalidated others through the manipulation of the research and the media (Healy, 2004).

The takeaway here is that treating mental health problems with drugs has caused as many problems as it has solved, if not more in the long run. The former director of the National Institutes of Mental Health acknowledged that four decades of drug development have not significantly reduced the morbidity or mortality of mental disorders despite the development of more than 20 antipsychotics and over 30 antidepressants (Insel, 2015). Close investigation reveals its failures in abundance and the tragedy that ensues. Yet despite this failure many in the research community have derogated NFB with minimal investigation and a reductionist bias that favors only one side of the arguments surrounding NFB research. There are no drugs that treat PTSD directly but only the symptoms of PTSD and these drugs have serious have long term consequences. In contrast, there is very good research in support of NFB including decades of successful clinical outcomes numbering in the hundreds of thousands. The most recent research finds NFB probably efficacious for both attentional issues and complex trauma (Larsen & Sherlin 2013; Nicholson et al., 2022). Our experience in the field of NFB is that we have a powerful alternative when used in conjunction with talk therapy especially when alpha theta (AT) NFB is used.

Chapter 2

Alpha Theta—A Brief History: The Other Half of Neurofeedback

Something New

Paralleling the ascendency of the drug solution in the treatment of disorder and the relative subjugation of talk therapy to the status of adjunctive solutions, was the gradual emergence of biofeedback and neurofeedback. This little-known approach was poorly funded with respect to research and derogated by almost every other related discipline. It was culturally and politically incorrect and worst of all, difficult to monetize. This difficulty led to a complete disinterest of corporate funding as well. Nevertheless, dedicated pioneers emerged and continued to develop the technology despite its ignoble reputation because it actually did work time and again in a very significant manner at the clinical level. The most common event that initially inspired clinicians to embrace it was direct personal experience of the reduction of psychological suffering for the clinician him or herself or for their family members. Most of these clinicians then imported it into their practice with stunning, and on occasion, almost miraculous results. The results seemed too good to be true to those not using the technology and often oversold by overly enthusiastic practitioners new to the field. Nevertheless, the clinical outcomes were real, enduring and documented in the form of endless case studies and case outcomes. The investigators of the original technology included researchers from UCLA, Princeton, and The University of Chicago just to name a few. The findings were valid, reliable and peer reviewed yet seemed too improbable to the scientific community and consequently discarded onto the "probably placebo bin" without serious review. Any scientist who came to that conclusion in reviewing research using animals and control group designs, as did Sterman's research, should revisit their methods classes and literature. The fact is that most detractors never seriously reviewed the research or were highly biased by their backgrounds and corporate connections.

The Development of the AT Paradigm

The history of the development of alpha theta training (AT) involved many players and to get the full story requires delving into many diverse sources, as each one contains only part of the story and only part of the players. The most well know source is A Symphony in the Brain, by Jim Robbins (2008), but the stories of other important contributors, such as Adam Crane and other east coast psychologists, are documented in the book Mindfitness Training (Crane & Soutar, 2000). Another important overlooked and out of print book detailing the very early development of AT, by the couple that invented it, is Beyond Biofeedback by Elmer and Alyce Green (1977). Additionally, a more recent source that fills in many missing details, is Alpha Theta in the 21st Century (Martins-Mauro & Kerson, 2017). To really grasp the history, you had to be there, and I had the good fortune to get to know and talk to most of the original major contributors. To understand the technical aspects of the "process," as Adam Crane referred to it, you also had to be reading the research and doing the clinical work with yourself and others. That was the most exciting part because of the immense changes it can often generate in people. This is especially important if you want to understand the most salient points in its evolution.

The story often begins with Eugene Peniston because he published the first high profile research piece formalizing the method in the later 80s. The real story, however, begins with Elmer and Alyce Green who did the original formative research and laid the foundations. Eugene had taken a workshop from Elmer Green and was inspired to use the EEG biofeedback techniques he encountered there with recovering alcoholics in his substance use disorder (SUD) program at Fort Lyons, Colorado. Green, a biopsychologist, and his wife Alyce, had begun working in 1965 at the Menninger Foundation in Topeka, Kansas with Dale Walters to develop a psychophysiology laboratory, which they named The Voluntary Controls Laboratory, to study muscle tension. They hoped to possibly learn to consciously control muscle tension through electronic feedback devices in a similar but more efficient manner than a progressive relaxation technique that was known at the time as autogenic training.

In 1966 they began a series of research projects progressively exploring different aspects of what was to be soon known as biofeedback. They began a series of in-depth investigations into how it could be used to control a wide range of physiological functions in the body and associated anxiety related symptomology such as headaches, migraines, insomnia, and ulcers. It is important to keep in mind that at this time voluntary control over involuntary functions like background muscle tension was considered impossible. Nevertheless, Elmer began building innovative methods of feedback for body temperature training and galvanic skin response (GSR) as well. He also became interested in the work of Lawrence Kubie (1943) a clinician and researcher who was inducing a state of "hypnogogic reverie" in clients with breathing sound feedback to gain access to unconscious information that proved to have a transformative effect. This combination of investigative interest soon leads naturally to the development of AT.

Elmer and Alyce found that gaining control of unconscious background muscle tension required a different form of focus than most people were familiar with and which was relatively unknown at the time. They had learned indirectly about this form of focus from doing yoga, also a highly unusual practice in the west at this time in history. They termed the ability to control this form of non-volitional focusing activity as "passive volition" rather than active volition. Adam Crane later named it "profound attention" (Crane & Soutar, 2000). Active volition does not work well in biofeedback. This is why, as I will explain in more detail later, using video fading during NFB sessions is so effective. With video fading the client does not focus directly on controlling the feedback but watches a screen that gets dark or light depending on the client's success while viewing a video. It conditions the client while their interest is focused on the content they are viewing.

They began training patients, referred by doctors at the clinic, in dealing with anxiety and phobias as well. In 1975 they began working on heart rate and blood pressure. They also worked with GI disorders, neuromuscular disorders and headaches. They had considerable success with most of the cases and pondered the limits of the technology. It occurred to them that any physiological system in the body could be harnessed, and even novel states of consciousness might be accessed through brainwave feedback.

In a meeting of the Psychophysiological Research Society in 1965, Elmer learned about alpha brainwave training from psychologist and sleep researcher Joe Kamiya who had discovered a method for training brainwaves and was doing research on this phenomenon at The University of Chicago. Kamiya (1969) had found that his subjects could actually learn to control a certain class of brainwave known as alpha rhythms. This excited Elmer because it suggested he might be able to duplicate the physiological states of relaxation and meditation by training brainwaves. His enthusiasm was further fueled in 1968 he met Barbara Brown at the same society and learned about her research on psychological state correlates of beta, alpha, and theta occipital rhythms. He also encountered the work of sleep researcher Barry Sterman, who in 1967 was beginning to investigate the brainwave frequencies known as the sensory motor rhythm (SMR). Barry discovered that SMR NFB training prevented seizure in cats exposed to a toxic rocket fuel while he was doing work for NASA at UCLA. Barry would go on to find that training humans with SMR significantly reduced seizure activity in epileptics (Sterman & Egner, 2006). Based on what he knew about biofeedback, these were impressive initial findings that held great promise for those investigating brainwave training.

The Greens began conducting research under an NIMH government grant in the 70s investigating the idea of enhancing creativity through brainwave training (Green & Green, 1977). Since the Greens were very interested in altered states of mind that led to enhanced creativity, they also took a great interest in meditation and its healing potential. Upon reading the research on Zen monks by Kasamatsu and Hirai (1969) and Anand, Chhina, and Sigh (1961) showing that deep meditative states correlated with theta, he began to focus on theta states in particular. Elmer had used a meditative state to solve difficult math problems in physics during his graduate school studies and was personally aware of what a powerful tool it could be. In reviewing the history of inventors and innovators the Greens had found that most of them had access to or consciously developed a method of getting into an altered state from which they derived critical insights. Alyce's research indicated that dreamlike theta states appeared to dominate the brain during episodes of great scientific discoveries such as the structure of benzene by August Kekiule, the periodic table by Dmitri Mendeleev, and the inventions of Edison and Tesla. Their goal was to develop technology to reproduce that highly creative state using EEG biofeedback and believed theta was a key indicator they were in that creative state.

Using his lab equipment, Green had previously determined that the meditation like state he had learned to access through yoga training and which had helped him solve math problems in graduate school seemed to be correlated with increased theta at a location designated as O1 in the international 10-20 system for electrode placement. This area was in the left occipital lobe. He also measured meditators who did a specific type of reverie style meditation and found their theta was very high as well. He encountered one meditator whose alpha even dropped down to 8.3 Hz, from the usual 9.5–10.5 Hz range that dominated most people's waking posterior region, during the meditational process. This finding was consistent with what Kasamatsu and Hirai had found when monitoring the EEG of Zen monks. A study by Dale Walters which was also indicating high theta during hypnogogic images further persuaded him he was on the right track.

Elmer began an exploratory project, prior to the official research project, training volunteer participants on a couch lying down in a dimly lit room using his EEG equipment which he operated from another room and using an intercom to communicate. Elmer focused progressively on uptraining theta frequencies when training his first subjects, but he discovered that it was especially difficult for his subjects to increase their theta. So, he began using a variety of pre-session exercises such a progressive relaxation, breathing exercises and hand temperature warming in an effort to achieve more success. He also found that those subjects who went too far into theta would not be aware of the images he was interested in stimulating, and developed an apparatus to keep them more alert that set off a door chime when they became too drowsy.

These initial participants came once a month for 7 months while practicing autogenic training and breathing exercises regularly at home. He continued to specifically focus on the alpha and theta frequencies as the primary goal of training. They trained for 30 minutes at a time with alpha and theta tones. He also reportedly experimented with training subjects until they were able to double their alpha production with their eyes closed and then train them in producing theta frequencies. After two sessions of training alpha on average, only the theta tones were presented as feedback as most subjects could easily produce alpha. His team asked the subjects what they were experiencing during each theta event they recorded during the experiments. Even those subjects who produced little theta initially were able to produce a fair amount of theta at this point in time. For Elmer, it became clear that theta was associated with images and that increasing theta until images occurred should be his goal as well as to do it in a manner that allowed subjects to remember the images.

The NIMH Research Part 1

The Greens then began their NIMH funded project to carry on their research and they began a two-part effort. Green designed and built his own NFB equipment for this NIMH project based upon his previous findings. The lab for his research project reportedly had a considerable amount of large electronic equipment, a reclining chair instead of a couch, and he consequently decided to use reclining chairs for the research. The equipment he built for the project, on the other hand, was relatively small and portable. They had not entirely refined the AT method at this point but had developed a pretty good format through the exploratory pre-research process. The NIMH research project was carried out at both the Menninger Clinic and Washburn University on 26 undergrads. They used the portable trainers to increase both alpha and theta regularly 1 hour daily for 5 weeks and then they would go to the lab to train every other week. Norris (2017) reports Washburn University devoted three rooms to the process and students signed in to schedule their time. They were provided with reclining chairs, clocks and journals and they were asked to keep notes after each session. During the every-other-week lab sessions, they were also hooked up to equipment to measure skin potential, heart rate, breath rate, skin temp, and galvanic skin response (GSR). Each session began with breathing techniques and autogenic relaxation. The subjects received alpha and theta training for 15 minutes and then theta training for 30 minutes. They ended each session with the subjects describing their experience.

The NIMH Research Part 2

Once the initial exploratory work was completed for the NIMH project, they began the second phase of their investigation. This was also a 5-week training period following the same procedures only they were supposed to focus on theta and encouraged to specifically report any hypnogogic imagery. They found it more difficult to generate theta and many began to fall asleep. We know today that this is a typical response to attempting to achieve these states without extensive meditative training as is traditionally done. With the greater focus on theta, images did increase and integrative experiences leading to psychological well-being began occurring. Some students had mild abreactions of fatigue and increased anxiety. Attempts to increase imagery unfortunately had the reverse effect and the experimenters concluded that a witness style effort worked best. The students were encouraged to just passively observe whatever occurred as much as possible. Imagery was coded on an adapted scale from Wallach and Kogan (1965). There was an unusual quality of quieting of body, emotions and thoughts that subjects encountered during their sessions. Major changes in the personal lives of the young students were not uncommon and reflected heightened creativity in terms of solving personal life problems. They also frequently experienced archetypal images such as a single eye, a wise old man, a tunnel filled with light, climbing up and down staircases, and a book of knowledge. This suggested they were tapping into states that earlier in human history inspired spiritual interpretations and were possibly the source of these them. Improved general emotional control was also noted and subjects often described it in terms of time to respond differently when triggered by an event (Norris, 2017). We will discuss the significance of this diminished reactivity pattern in later chapters at more length.

The abundance of symbolic and archetypal imagery that occurred during the experimental sessions also lead them to an increasing interest in experimenting with guided imagery. Pat Norris, who was working with them at this time, had received a Ph.D. in this area, and she began implementing techniques she had acquired in various sessions and then more extensively in other projects later. She found that in many cases utilizing guided imagery with clients significantly accelerated their progress, especially when they had difficulties attaining the crossover experience. When initially experimenting with autogenic training, and then with biofeedback, as mentioned, Elmer found that willful attention was not effective and that imagining and visualizing had a greater effect on the body. Ironically, he had learned the basics of visualization to alter body function from a gentleman named Arthur Green, who in turn had learned it as a student of the Sufi tradition rather than from the limited psychological literature on the topic. This was part of what he defined as "passive volition." Visualization was later to be utilized further in Peniston's procedures and was eventually considered a critical component of the process. My wife Barbara and I later advocated extensive use of visualization, guided imagery, dreams, and symbolic interpretation based on our understanding of the work of Carl Jung and

Joseph Campbell as well as Barbara's training in shamanistic methods of healing. We have found that these methods greatly enhance the process overall and we have continued to develop them and incorporate them in our workshops. Today visualization training is used extensively by professional athletes and its value is well understood regardless of the blatant minimization and rejection of it by the medical community. This situation, however, is slowly changing.

As one can see, the effort, time, and money that went into the development of this technology was considerable. It was based on research developments throughout the twentieth century and was supported by government grants. It was well executed and well thought out, but it was very innovative and impregnated with ideas that were very unpopular with the reductionist and materialist scientific paradigm that reigned supreme at the time. It was generally ignored or rejected by the mainstream, as new and innovative ideas often are, and received minimal funding from this point on. Historically, the same narrow minded collective of scientists blocked many other new innovations of the time because they threatened the status quo. This as it turns out is a very normal part of modern science (Kuhn, 1970). Eventually innovation resurfaces and is validated when the old guard passes. One investigator, fortunately, managed to validate it and frame it a more acceptable light, although even that made a limited impression on the scientific community of the time.

The Peniston Formalization of the Method

In addition to the NIMH funded work there were also many opportunities to try the technology on cases of hospitalized mental patients referred by physicians at Menninger to the Voluntary Controls Laboratory that Elmer developed. These cases for the most part had successful outcomes, and this led to further interest in a more formalized research program (Norris, 2017). For this population the MMPI was used and the first three scales, Hypochondriasis, Depression and Hysteria, changed very little but Paranoia, Psychasthenia and Schizophrenia changed by as much as 2 standard deviations after training. There was, however, a little backsliding after a 3-month follow up (Norris, 2017). It was at this time Norris had started using guided imagery, as mentioned previously, with patients as well and in one client with almost no alpha and alpha increased significantly. This indicated that guided imagery could be very useful at helping those with low amplitude alpha and difficulty engaging in alpha theta training.

Not only are there alpha and theta brainwaves but there are beta brainwaves, as well as others which will be discussed in detail in the next chapter. Further experiments of randomly training either beta brainwaves or alpha brainwaves separately showed that all patients benefited from beta training. Patients reported feeling more present. Based on what we know now, that was likely because many of them experienced excessive slow brainwave activity and the increase in the faster beta frequencies reduced the presence of the slower brainwaves and helped them with frontal lobe function and cognitive processing. This finding would have ramifications for further developments later in the field, especially the work of Sue and Siegfried Othmer and Joel Lubar with regard to ADHD. Some of the uptraining of beta frequencies may have been inspired by discussions at the meetings they attended regarding Sterman's work at UCLA with SMR training, which is technically in the beta frequency range.

Workshops for other professionals began to develop out of Elmer's explorative work as well. Dale Walters and Pat Norris, who were affiliated with Green's experiments, began training substance abuse patients from the hospital as well. Other studies had shown alcoholics who had gone through withdrawal had almost no alpha, but once they consumed alcohol, their alpha returned. This has also been found in many other subsequent studies (Norris, 2014). Peniston was working at the time with addicts at Fort Lyons, CO using a conventional substance use disorder (SUD) program, and attended one of the workshops. Upon learning about the alpha studies from Walters and Norris, he became very interested in the possibility of increasing the alpha in alcoholics to help them in treatment. This resulted in his utilization of AT training in his own program with 10 of the subjects doing AT and 20 others in two control groups (Scott, 2017). He followed the Menninger protocol as closely as possible. The outcomes were impressive and published in a peer reviewed journal. The Menninger Clinic soon became the Center for Applied Psychophysiology and Biofeedback and then in 1993 spun off the Life Science Institute of Mind-Body Health. Studies with successful outcomes continued to develop. Norris and colleagues concluded from their further research with addicts that booster sessions after one year would likely be helpful. The idea of booster sessions was an entirely new concept at this point in time, and would eventually become a standard practice as clinicians discovered that further stress and trauma experienced by clients could cause post-training backsliding. It is not unusual to have present day clients go through successive rounds of NFB over a period of years as they work through different layers of their complex trauma.

When Peniston took the aforementioned workshops at the Menninger clinic he had a vision that he could use this technology on alcoholics to assist them in recovery (Robbins, 2008). Peniston then later published in his notes, citing influences from Vogel, Foulkes, and Trosman (1966), that he observed that as individuals became drowsy, their brain waves commonly changed from high-amplitude alpha to low-amplitude theta. He further notes that the transition from alpha to theta produced a unique hypnogogic state.

During the transition, some individuals experienced a hypnogogic state in which they had vivid visual imagery and auditory and visual hallucinations. Investigators studying creative individuals noted that when their subjects were in a state of "reverie," they produced increased amounts of 6–8.5 Hz activity (Green, Green, & Walters, 1974). In an effort to facilitate production of the reverie state and hypnogogic imagery, the investigators developed an alpha-theta biofeedback system that provided information to the subject about both alpha and theta production (Peniston, 1998).

Peniston had apparently been aware of the influence of alpha training on alcoholics who tended to have low amplitude alpha by investigators such as Passini et al. (1977) and others but also aware that Passini was concerned that changes in alcoholics due to alpha training may be due to a placebo effect.

Peniston, along with his co-author Paul Kulkosky, then deployed the protocol we use presently in his own clinical setting. This first study divided 30 subjects into 3 equal groups of 10 subjects with an AT group, a group that received conventional treatment and a non-alcoholic group of hospital staff (Scott, 2017). They used a variety of well-established pre-post psychometrics on all three groups to measure outcomes as well as measuring beta-endorphin levels in the blood. He pretrained patients with biofeedback, relaxation and visualization exercises, like Green had done, and used an abstinence visualization at the outset of each session. Each patient would visualize themselves being abstinent in some manner particularly relevant to their life situation.

Pre-post measures from the experiment showed reduction in Beck Depression Inventory (BDI) scores and lower levels of beta endorphins and less relapse over 13 months than controls. They found reductions in over 13 scales on the Millon Clinical Multiaxial Inventory while controls showed only reductions in two scales. The AT group also showed improvement on seven scales of the 16-PF compared to one scale on controls. So, changes in psychological and physiological measures were both established. There was 80% abstinence with the AT group vs 20% abstinence with controls in a 4-year follow up. Soon, in another series of experiments, Peniston and Kulkosky published findings on experiments with both alcoholics and veterans with PTSD in the late 1980s. These experiments briefly generated considerable interest and excitement because of their profoundly successful outcomes (Peniston & Kulkosky, 1991). Unfortunately, his initial studies were not always exactly clear in the methodology section, written by his graduate students, and this continued to generate confusion with regard to replication as the years passed. This unfortunately led to confusion regarding how to implement the protocols clinically and how to replicate the study.

Similar results using AT training followed with the Saxby and Peniston study (1995), the Johnson and Bodenhamer-Davis study (2017), Callaway and Bodenhamer-Davis study (2008), and the Kelly study (1997). The main complaints regarding all these studies by other investigators were the use of multiple treatments involved, such as other forms of feedback, breathing exercises, and visualization, and confounding the effect of AT. There were also issues as well as the small subject sample sizes they usually employed. While technically valid methodological complaints, they tended to overlook the importance of the overall effects of the entire protocol. The other treatments together had never previously produced such results without AT training and vice versa. The bio-psycho-social nature of these innovative methods involving biofeedback were revolutionary for their time. It must be recognized, however, that AT training is still not a stand-alone method for dealing with complex trauma and still requires adjunctive methods even for peak performance training with relatively healthy individuals.

Bill Scott and CRI-Help

Peniston's experiments inspired a long series of research efforts, especially by Bill Scott, that culminated in the CRI-Help study done by Bill at UCLA department of Psychiatry (Scott et al., 2005) as well as other studies abroad (Egner & Gruzelier, 2003). Bill had was working closely with Peniston between 1993 and 1995 conducting a neurofeedback investigation at the Northlands Recovery Center addiction treatment facility in Grand Rapids, MN. He was having difficulty interpreting and replicating the methods section in Peniston's paper and called Eugene for his help. It was a small study with 24 Native Americans and Bill was able replicate the protocol with Peniston's help with one exception. Bill was troubled to find that subjects with alpha greater than 12 μV became more anxious when training alpha up. Most of Peniston's small sample had low alpha amplitudes and apparently never encountered this problem. So, Bill innovated and trained alpha down initially instead of up, which surprisingly reduced their anxiety. Today we know that high amplitude frontal alpha is not uncommon in individuals with general anxiety disorder (GAD; Kerson et al., 2009), or low alpha with complex trauma (Ros et al., 2013), but back then investigators were assuming higher alpha always meant greater relaxation. I will cover this issue in more detail in the chapter on the New Mind Alpha Theta protocols. Bill, to his credit, in more recent times also notes in his account of the research that people with PTSD have diminished posterior alpha, citing Jokic-Begic and Begic (2003) and Wahbeh and Oken (2013).

Bill also made some other innovations during this investigation. He found that many of the subjects were falling asleep, so he inhibited delta to reduce tendencies to go to sleep and found it seemed to also reduce abreactions in all the subjects. He was skeptical at the time that everyone with a history of emotional trauma would likely have negative reactions as part of their healing process (based on my discussions with Bill) as Peniston had warned him (Scott et al., 2017). I and many other therapists who worked with alcoholics and 12 step programs agreed with Peniston on this, but many therapists were looking for a pain-free method to manage PTSD. I discussed this aspect of the training with Peniston's co-author Kulkosky and he confirmed their experience during their research with Vietnam vets (Peniston & Kulkosky, 1991). Integration for this population was a painful process, so much so that Kulkosky would have called a halt to the experiments but for Peniston's clinical expertise and skill. Contemporary research on complex trauma supports this observation as well (van der Kolk, 2014; Scott et al., 2017). On the other hand, Bill was legitimately concerned about dealing with such events in such an experimental protocol and he was also concerned subjects might stop showing up, which does happen sometimes. Another innovation Bill initiated at this time was to use Pz instead of O1 as a sensor placement point so that the sensors would not always fall off, but later added because of an article he read in Scientific American by Joseph LeDoux that indicated there were important connections with the amygdala in that region (Scott, personal communication, 2002).

Bill performed a 1.5-year follow-up and confirmed a 79% abstinence rate, which was unprecedented in the field of addictions, except for Peniston's work. All subjects had met DSM III criteria for PTSD. Unfortunately, this study was never published, according to Bill, because of methodological issues. But this led to protocol modifications that many others began to imitate at the clinical level. After over a decade of clinical experience and research Bill went on to do one of the most important research projects to date in the field of alpha theta training known as the "Cry Help" study.

The "Cry Help" study was named after a clinic where it was conducted at the CRI-Help, Inc residential treatment center in Los Angeles and involved 121 mixed substance abuse volunteers. The outcome was very impressive. Subjects were intensively trained with eyes open fast wave protocols twice a day for 14 days. Most showed statistically significant improvements on their Tests of Variables of Attention (TOVA) in impulsivity and variability but only modest improvement that was not significant in attention (not surprising since trauma had not been resolved). They were then trained for 30 sessions with AT training. Most received a total of 50–60 sessions. Bill had pre-post MMPIs done on the subjects as well showing significant improvements in 7 out of 10 scales. Again, like Peniston's experiments, the outcome was unusually impressive but with a much larger sample size. Most of the subjects who finished the experiment recovered with minimal relapse and 77% were still abstinent after 12 months compared to 44% of controls. Our present clinical model of first fast and then slow wave training still retains many aspects of this research and reflects the methods employed.

Using the Scott-Kaiser modifications of Peniston, Dehghani-Arani et al (2013) replicated these results with 100 opiate addicts and Rostami and Dehghani-Arani (2015) did the same with amphetamine addicts. Rates of PTSD with substance abuse are approximately 50% (Brady et al., 2000). Cleary AT is effective with PTSD. Of significant importance was the finding by Scott-Kaiser that a 12-step affiliation was critical for higher patient retention. This was the model we had been introduced to in the early 1990s working with my research advisors, Doug and Marianne Frans, at the University of Arkansas where they had set up the first clinic in the state training alcoholics and clients with PTSD. At the time we were running 12-step study groups as an innovation for emotional trauma, and it was clear that the AT component added a critical dimension to the process. Doug Frans, who was head of the Social Work Department at U of A, and I were investigating the relationship between family systems and alcoholism in our research, and I was impressed by the impact that AT had on the group meetings he was holding at the clinic. We were both convinced that a major component of alcoholism was emotional trauma, especially arising in the family system.

Another thing Scott uncovered during his research was the finding that those with opiate addictions will have trouble with crossover. I too found from over decades of clinical qEEGs, that opiates generate stubborn elevated slow frequency alpha and not only alpha theta training, but most forms of NFB are useless in these situations, as the opioids force the brain into a rigid pattern of EEG of high amplitude alpha. This pattern is very unresponsive to any effort to train it unless the opiate is withdrawn, but then for many, the resulting pain that the opiate helped them manage equally interferes with training. It's a catch-22. Another important conclusion he reported was that fast wave training prior to doing AT training actually enhanced AT outcomes, especially frontal slowing associated with ADHD. The rationale for this was also detailed in Sokhadze et al. (2008). Some further important refinements to come out of this study that inform our present methods are:

- Clients may require booster sessions every year.
- Clients can train everyday twice a day and achieve faster results.
- Further confirmation that all patients seem to benefit from beta and SMR training initially before AT, especially those with frontal slowing.

- Adding guided imagery can assist clients in achieving better outcomes.
- Gentle attempts to rouse sleepy clients work best.

Gruzelier and Egner Studies

Egner and Gruzelier (2003) conducted research with some guidance from Bill Scott at the Imperial College of London School of Music using AT training to enhance musical performance. Their study showed significant improvements in the overall quality of performing musicians including instrumental competence, musicality and communication. These encompassed all aspects of musicality such as stylistic accuracy, interpretative imagination, expressive range, emotional commitment and conviction, deportment, and ability to cope with the stress of formal recital. Gruzelier continued research in this area demonstrating the enhancement of creativity in general and in support of the Green's findings. Gruzelier recites nearly two decades of research by himself and others in multiples areas of performance enhancement using AT training. These include ballroom dancing, singing, and even medical student's confidence related to performance.

Egner et al. (2002) conducted an experiment demonstrating that the effects of AT were not due to relaxation alone. This was a critical piece of work because most of the previous experiments were confounded by other methods. He set up an experiment to compare EEG signatures between AT subjects and mock control subjects randomly assigned. He used a 2-minute baseline and a 50% threshold for both alpha and theta to set thresholds. Both bands were rewarded as long as they remained in a range of 30–65% above thresholds (a bizarre method). He used only 15-minute sessions (clinically not long enough for many first timers to achieve crossover). They did only five sessions! They did use same time of day for each session.

They averaged EEG in 3-minute blocks—so they had five averages. They found AT group did indeed increase AT ratios compared to controls in each of the latter parts of each session. They tended to increase AT ratios across sessions, but findings were ambiguous. They acknowledged many possible confounds for across session findings. Interestingly, the AT group did not subjcctively note significant changes as compared to the controls. I would argue the session were too short and there were not enough of them to make a difference.

Gruzelier continues up to the present time to conduct investigations demonstrating the efficacy of AT for a wide variety of neurological functions. Although this is not a book on peak performance, the findings from the studies support the efficacy of AT training, especially with regard to creativity. Later in the book we will discuss why the spontaneous emission of novel and creative behavior is so critical to the integration of trauma.

Chapter 3

The Development of The Method and Theory: More Details

Basic Concepts in Neurofeedback

In order to better understand and interpret the chapters that follow it is important to have some basic understanding of standard neurofeedback, the basics of signal processing and how neurofeedback is done. This may seem challenging and a good time to skip a chapter, but I will make it as easy and short a discussion as possible. It is only necessary to understand the basic ideas and concepts and not the details unless you plan to become proficient at it.

Brainwaves are very faint by the time they reach the scalp, and they must be amplified in order to be processed by a computer. The amplifiers for brainwaves come in very small boxes with ports for connecting them to your scalp with wires and ports for connecting them to your USB port on your computer. The wires are attached to the scalp using a tacky paste that conducts brainwave electrical activity into the wires that are running to the amplifier. Once the brainwaves are amplified in the amplifier box, they are also converted to digital information for the computer to process. The computer in turn slices and dices them into different frequency bands. Much in the manner that a prism breaks light down into color bands of different light frequencies, computers break the basic brainwave signal from your brain into different frequency bands. These bands are named, from lowest in frequency to highest, as delta, theta, alpha and beta, rather than colors of red, blue and green. Delta is the lowest frequency band and is typically related to sleep, while theta is more related to memory functions and emotional processing. Alpha tends to be involved in resource management coordinating different regions of the brain, so they work harmoniously together. Beta is the high frequency hard worker that does the cognitive processing, especially planning and coordinating behavior. This is a very simplistic representation but sufficient for its purpose.

When these frequencies appear in the right amounts and in the places, all is well and they reflect a person functioning harmoniously with his or her social context and identity group. Excessive stress, accidents, toxins, and diseases can, unfortunately, disrupt this harmonious order in both directions and result in both mental disorder and behavioral disorder which can be observed as brainwave disorder as well. We can see this disorder in the mapping process of these brainwaves as well as in the trend screens we use to monitor them while training. It is possible, as demonstrated by Barry Sterman (Sterman & Egner, 2006), to train and reorder these brainwaves in both animals and people to reduce the disorder from various stressors and help return the brain and central nervous system to better order and its reflected functioning. In fact, Barry was able to use neurofeedback to reduce or eliminate the natural seizure activity that occurs in cats when exposed to hydrazine rocket fuel. He then went on to reduce seizure activity in people that were nonresponsive to seizure medication.

It goes without saying that this upset the scientific applecart in the twentieth century, and everyone did their best to try to ignore or derogate Barry's research. This is difficult to do to a very talented and dedicated anal retentive investigator at UCLA with a grant from NASA who was using airtight control group designs with animals—no placebo effect thank you very much. Nevertheless, it is an old story in science and medicine that innovators going back to Galileo who radically advance our understanding are sure to be labeled dark and misguided characters undermining the reasonable order of things. The concept of brainwave training has been shunned by "reasonable" scientists ever since. The unfortunate problem is that it does work and is growing in popularity just like aspirin did, even though no one knew how it worked. But I digress.

Today well-designed neurofeedback research is rapidly growing around the world and there are hundreds of studies to date. Today we also have sophisticated software that tracks these brainwaves as we train people to change them. We use something called involuntary operant conditioning to do this in most cases. Sensors on the scalp send information to the amplifier and then the computer and the software is set to increase or decrease sound or the contrast to a video, so it fades or gets brighter when the frequencies we want to influence increase or decrease in amplitude. If we want one frequency, such as alpha, to decrease in amplitude, then whenever it goes down the client can see the video more clearly and when it goes up the client gets a screen so dark, they cannot clearly see the video. By just watching the video, their brain gets trained to decrease alpha.

Over time, after many sessions, the alpha will usually tend to remain lower in amplitude and clients will have less of the mind chatter frequently associated with having high amplitude alpha all of the time. In addition, they will feel more calm and centered. We can also train beta up or theta down in many cases to improve attention. There are many possibilities and the purpose of mapping the brain beforehand is to identify areas that are statistically outside the usual normative patterns of activity. The brain maps, or quantitative EEG maps (qEEG), help clinicians decide on the best training strategy. This type of brainwave training is often traditionally referred to as "fast wave training." It is the initial type of training we use to emotionally stabilize people with trauma and help them sleep better before working directly with the trauma through AT training.

In fast wave training, we are usually attempting to permanently normalize a specific location or region of the brain. With AT training, the emphasis is using brainwave feedback as a means to guide clients into a specific liminal state between waking and sleep. In that sense it is a bit like hypnosis but a different process for achieving an alternative state of consciousness.

It is in this liminal state where clients are able to recall memories and feelings related to trauma and to integrate them.

This may all sound exotic and techy but most people with average computer skills can easily learn to do it. The more difficult aspect is interpreting the brain maps and interpreting the training screens. New Mind has spent a decade making this part relatively automated so that the map interpretation involves just reading the auto report which interprets the map for you. Interpreting the trend screen training results comes down to watching a few key numbers. The AT training, on the other hand, requires more practice reading the trend screens but we will provide a significant amount of detailed guidance in the chapters that follow. Most clinicians go through a mentoring process where they receive this guidance and can ask questions as well as provide their own client examples for analysis.

The Basic Protocol Format: Acquiring the Liminal State

I think it is important to provide an overall schema with regard to AT training at the outset. The method, as we have said, was originally devised by Elmer Green with additions by Pat Norris and Dale Walters. It evolved under their experimentation and eventually was adapted by Gene Peniston for his research projects. The form it finally took was developed between Bill Scott and Peniston as Bill continued to do research. Finally, there were the Scott-Kaiser modifications from the research done involving David Kaiser and Bill Scott on the CRI-HELP project. Peniston was very clear that AT was not a standalone approach for him but an adjunctive technique to an overall method (Peniston & Kulkosky, 1999). In fact, he failed to mention that it was done in conjunction with a 12-step program. This turned out to be a critical omission with respect to how it was perceived by many clinicians, and even some researchers. Some attempted to employ it clinically to clients with addictions without the necessary group therapy or a 12-step program (Scott, 2017). Although people have frequently tried to just "cherry pick" the protocol, they are technically not doing AT when they use it alone without the other associated techniques as I will be discussing in later chapters, and they are ignoring a critical aspect of the research. Consequently, they often may not get the same results.

Peniston developed a preparatory regimen prior to AT training because he saw from Green's work it was necessary to assist people in developing the skills needed to achieve the deep states potentially available with neurofeedback. This involved progressive relaxation practice, visualization practice, the biofeedback practice of hand warming and breath control methods. Once clients were accomplished in all of these areas, they would then begin the AT training. They were placed in a silent environment in an isolation room with a reclining chair and with a single window used to monitor client status. They were hooked up to the neurofeedback equipment and began the session with their visualization. The patient was then instructed to listen to the audio tones, one representing alpha wave production and the other theta wave production. They were left alone in the room for 30 minutes of training listening to the tones. Afterwards the client's experience was briefly reviewed, especially with regard to any images and then they were done with the session. In addition to this procedure, they also usually went to some form of group counseling, which was later identified as a 12-step program, and discussed their experiences.

Having taught this method for a couple of decades, I have frequently encountered many clinicians who think they are doing AT but who are unaware of the details associated with this procedure beyond just hooking clients up and giving them 30 minutes of feedback. In most of these cases, they would just pick the protocol from the manufacturers list of available protocols to load, run the client for the required amount of time, and proudly proclaim, and vehemently argue, that they have been doing AT training. Even worse, they used auto-thresholding during the process which actually defeats the whole AT strategy. Yes, they may be doing neurofeedback and clients may be having some positive experiences, but it is technically not AT training. It should be obvious at this point that AT training is multimodal and complex. When they actually learn to do it properly clinicians are surprised to find out how much more powerful the method is and how much more effective they are when using it.

Much has happened in the way of research in the neurosciences since these seminal studies in AT were conducted. Good explanations for specific aspects of the method were often never discussed in sufficient detail to satisfy most conscientious practitioners. There were lots of conflicting theories and details surrounding the procedure. In spite of this, I have been able to find explanations for some of the mechanisms of action of the process of AT with the help of further research over the last two decades and drawing related sciences that contains information not previously available. My wife Barbara and I have evolved Elmer Green's and Peniston's methods, especially the clinical components, over a period of 20 years through clinical application and experimentation. I have developed software with my development team at New Mind that standardizes the technical side of the process and made it safe and simple to implement. We have presented our discoveries and methods in workshops at Future Health, The Association for Applied Psychophysiology (AAPB) and The International Society for Neurofeedback Research (ISNR) over the years since 1999 and continue to do so to date.

The general interest in AT in both the field of neurofeedback and counseling has, nevertheless, been modest and varied. I suspect it is because it requires a variety of skills not readily available to the average clinician. To be proficient in the methods also requires the clinician to do their own training before doing it with others and this unfortunately is not always done. I think this may be the biggest hurdle for most trainers. In addition, there is the cultural drive for the quick fix and for painless resolution of symptomology which is not realistic with complex trauma. However, there is emerging renewed interest in AT as other methods, as effective as they are, fail to fully resolve trauma or even resolve trauma at the same level as typically occurs with AT. More and more practitioners are also running into a roadblock in their general NFB training methods with "difficult clients" who seem less responsive. These individuals with severe trauma often fail to continue to make progress and resolve their trauma. This is because they are missing the other half of the NFB method known as AT training. We at New Mind believe that we have evolved the AT method into a sophisticated approach, more closely allied with Green's original interests. This approach takes individuals not only into their unconscious but deeply into their collective unconscious as well as their spirit and compels them to work creatively with their life and transform it. In that sense, it is more along the lines of 12-step methods. Not every client is ready for this process but in our decades of clinical experience and mentoring we have found this to be one of the most efficient approaches to resolving complex trauma. The new research in psychedelics is also supportive of this broader and deeper perspective that incorporates the qualia of experience and not just data driven objectives and the concepts that are derived from the reductionistic and exclusively materialist quantitative paradigm.

The Theory

Elmer and Alyce Green developed the original theory and technique of AT which was outlined in their (1977) book Beyond Biofeedback. It grew from a more wholistic and esoteric paradigm and was antiseptically adapted for clinical use in a manner that would not offend 20th century mainstream psychology, and the policies that were derived from that paradigm, to please the economic players who determined the funding for any research and development. Behaviorism was decaying and the sixties had opened some frontiers but a collective western fear of qualia and the obsession with a reductionist agenda still held sway. Psychology was still trying to prove it was a deductive science that utilized control group designs and statistical analysis to do research, and psychiatry was just discovering psychotropic drugs and the professional power and status they brought to the profession. Without Peniston's careful implementation and characterization of the method and without the control group design and publication of his work, AT would probably have faded into history as another passing clinical fad—more of the touchy-feely wu-wu stuff that clinicians were always fooling around with. The "serious" clinical investigators were developing cognitive behavioral therapy (CBT) using "proper designs" and statistical approaches and drawing from a carefully vetted cache of research.

Elmer's perspective and method was still burdened with too much unsupported theory from a reductionist perspective, and was somewhat embarrassing to the research community. He also used many case studies in his exploratory research, and this was something mainstream psychology and psychiatry would, at that time in history, rather leave behind. In addition, he incorporated many ideas from a wide variety of metaphysical schools of thought, and it was his serious investigation of them in addition to what he learned at NASA that resulted in major contributions leading to the development of the field of biofeedback. Today we are becoming more aware of the problems with the application of purely reductionist methods to human behavior (Hoffman, 2019; Soutar, 1996). New methods combining both quantitative and qualitative methods, as Green did, are emerging in serious science today. Scientism, the cult of reductionism, is on the wane.

As a young man, Green had learned the basics for the method from an "Irish Yogi" who had taught him meditation and concentration exercises. It is important to interject here that today we know there are many types of meditative techniques, and they result in different qEEG configurations (Davidson et al., 2003). Green began to apply this particular meditative technique in college to help him solve math problems that he had difficulty with. Later, when he went to India in 1973, he found that others knew of the technique, and he discovered the same EEG pattern in those individuals that had appeared in his brain when he engaged in the method. He theorized that this was a liminal state between waking and sleep that allowed for creative solutions to emerge. One of his psychiatrist friends referred to the particular EEG biofeedback method he developed as "dredging the unconscious," and Elmer agreed this was partially what he was doing. One of the key measurement features of this state was high levels of theta over the area of the brain referred to as O1 in the 10-20 coordinate system. That is all he knew at the time because qEEG and the associated neurometric dimensions of measurement associated with it had not been developed and defined.

A Liminal or Twilight State

It requires years of practice of meditation in order to develop proficiency. One does not just sit down and master the technique in a day or a week or a month. This is a western fantasy that is constantly renewed in the spiritual marketplace in a culture heavily immersed in spiritual materialism (Trungpa, 2002). Elmer found that he could create a sort of electronic short cut to achieve a momentary experience of some of the states that are a by-product of the journey to mastery of meditation. Nash (2017) recounts his first Peniston workshop where an Indian Yogi was trying out AT and after 30 minutes he exclaimed "Just like yoga—only faster!" Green had very developed theories on this idea. Again, most of these ideas and theories have been ignored or forgotten in the "cherry-picking" frenzy that has followed, in the same manner that the other aspects of AT training had been ignored in the Peniston protocol. Even Peniston was uncomfortable with Elmer's esoteric interests and did not incorporate that perspective into his clinical model. Had he done so, he would likely never have gotten the research published. This cherry-picking tendency is also reflective of the effort by westerners to focus on meditation mechanics and ignore the preparatory training and theory associated with it in eastern culture and which is considered crucial to a successful meditation practice. Hatha yoga was primarily developed to prepare the body for meditation, not for the purpose of having a beautiful and healthy body as is often the case in modern times. Many clinical neurofeedback practitioners today avoid focusing on breath training or other biofeedback modalities such as temperature training that help prepare the client for AT training. In fairness, Bill Scott did not use preliminary training for AT in the CRI-Help study, although 12-step and personal counseling was part of the CRI-Help study, and he did get very good results. However, our three decades of clinical experience and experimentation using hundreds of clinics has taught us that better results could likely have been achieved using a more comprehensive clinical model.

It should also be acknowledged, to be fair, that Elmer's concept of training was, to some extent, also part of our modern cultural drive to strip away everything but the core method, "Just like yoga—Only faster!" He wanted to develop an electronic shortcut to these deep meditative states that was available to the common man. In Beyond Biofeedback, Green & Green (1977) states: ". . . my colleagues and I (Alyce Green and Dale Walters) eventually succeeded in instrumenting and generalizing the meditative method through alpha-theta brainwave feedback training. And we subsequently made it available to everyone, meditator and non-meditator."

He began his initial efforts to do this by training theta but also by training alpha first because it was so difficult for people to access the theta state. During his initial research, Green found that after a few sessions most people were able to get into the theta state for several minutes and "interrogate the unconscious for the solution to problems," plant visualizations, or explore higher states of consciousness.

Individuals training in these states during his experiments often reported intense dream-like experiences and visions that were frequently similar in content to life after death experiences. On many occasions these bled over into real life events and seemed to influence or become a part of daily life. His goal was to "train them in voluntary control of certain existential states whose central-nervous-system correlates are revealed by the presence of low-frequency alpha and theta rhythms in the EEG record" (Green & Green, 1977 p. 17). Note the inclusion of the term "low-frequency alpha." Alpha is broadly defined by the aggregate frequencies of 8 Hz, 9 Hz, 10 Hz, 11 Hz and 12 Hz and most people have more power and higher amplitudes in the 9 Hz–11 Hz range. So low frequency alpha would be alpha < 9 Hz. Today, we know this is true in one type of meditation known as concentrative meditation (Dunn et al., 1999). Maxwell Cade (Cade & Coxhead, 1987) who was also investigating altered states and meditation also later reported in his research that more advanced meditators he measured had elevated alpha frequencies below 9 Hz. Elmer further identified the state as "a poised non-drowsy state" that "appears to facilitate recall processes."

Elmer did not see himself as training people in brain-wave control or operant conditioning as is commonly proposed for fast wave training that is typically pursued for remediation of symptoms in the contemporary neurofeedback paradigm. He was very specific regarding what he was attempting to do and what he thought was taking place from a scientific perspective as a biophysicist. He was trying to duplicate the physiological state of the brain that correlates with the meditative state he was interested in. It was a control systems approach that he learned in his work at NASA.

"In actuality there is no such thing as training in brain-wave control; there is training only in the elicitation of certain subjective states that are accompanied by oscillating voltages in the cerebral cortex, detected through the subject's skull and scalp. Brain waves, as such, are not known to have any sensory representation whatsoever by means of which they can be detected. What are detected and manipulated in some unknown ways are foci of attention, thought processes, and subjective feelings (Green & Green, 1977)."

So, where did the alpha component of AT training come from? Elmer is quite clear on this as well. He trained alpha because it was an easier frequency to teach people to influence initially. In Beyond Biofeedback (1977) he specifically states that in his preliminary study he began to train theta only after the first two sessions, "because the subjects had little difficulty in increasing their percentage of alpha and we were particularly interested in investigating theta" (Green & Green, 1977, p. 137).

During his initial research, Green found that after a few sessions most people were able to get into the theta state for several minutes and, as I have said, "interrogate the unconscious for the solution to problems," plant visualizations, or explore higher states of consciousness. In this theta state his subjects not only discovered solutions to problems but also experienced other features of the state which were "hypnagogic and dream-like images during a state of semiconscious reverie... in conjunction with periods of theta rhythm and low frequency alpha waves" (Green & Green, 1977, p. 10). It can be deduced from this statement that it appears that Green was not entirely clear whether the correlates of state were theta or alpha as he suggested training in "theta waves or low frequency alpha (Green & Green, 1977, p. 16)" or a combination of the two but he clearly felt the brain could find its own way once it was guided in that direction.

Elmer developed a comprehensive theory of consciousness which he outlined and presented at various conferences. Below is a map of consciousness he used.

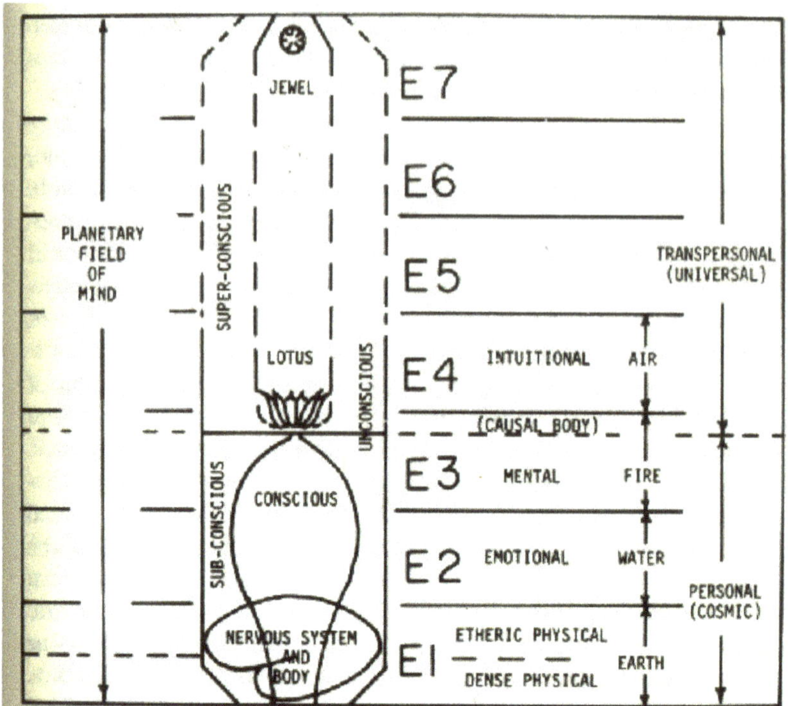

Figure 17. **Symbolic interpretation of human substance and perceptual structure.** Each line of the vertical cylindrical figure, which represents a human being, stands for at least three things: a boundary between different organizations of *substance* (different kinds of matter), a boundary between different kinds of possible *perception*, and a boundary between different kinds of possible *action*.

The Alpha Theta Environment

Elmer trained subjects in an isolated room with dim lights (as discussed in a previous chapter) and at times a window to observe the subject and an intercom to communicate. Subjects would lie down on a couch or reclining chair during the sessions. The room was mostly empty with minimal distractions and silence was maintained during the session. One exception on early investigations was a doorbell that was installed to wake clients when they became so drowsy that they entered stage 1 sleep. He tied the doorbell cord to a ring on their finger and had them keep their hand elevated from the elbow up so that as they became drowsy, their hand would drop pulling on the cord and setting off the doorbell. It was similar to Edison holding ball bearings in his hand as he drifted off. Elmer felt it was critical to avoid any distractions or intrusions into the session. In his later research project, he began using other methods that were less intrusive. Based on his research it can be surmised that a comfortable and quiet room with dim lighting is crucial to the AT process. Many clinicians use a recliner, but this can be a problem if clients are sleep deprived. They tend to fall asleep more easily. Bill Scott would prefer the method of asking his subjects to move a hand or a foot when they appeared to be falling asleep in order to rouse them back to conscious awareness. This, however, was also after trying to achieve the same goal with a tone that was meant to wake them when their delta went too high.

Alternative Techniques

Green tried many approaches to brainwave training. He trained people extensively with alpha and then theta initially. This was because he felt that the alert calm state generated with alpha training would assist them in maintaining conscious awareness when they entered deeper states close to sleep. He often did this by training both alpha and theta at the same time. Eventually he focused more quickly on getting them to enter the "theta state." The initial technique, as mentioned, was to first train one frequency, alpha, and then the other, theta. He also experimented with training alpha frequency down to 8 Hz or lower instead of amplitude training. So, Elmer would train subjects to lower their dominant alpha frequency until it actually moved into the theta range but decided amplitude achieved better results. This slowing of alpha in meditators had been known since the publication of Kasamatsu and Hirai (1969) wherein they studied the EEG of Zen meditators, and Elmer specifically refers to this study in his book. Elmer would train subjects to lower their alpha frequency until it moved into the theta range. This, of course, confounds his experiments to some degree, resulting in some confusion over whether he was training clients into a state of slowed alpha or theta. Unfortunately, there are those clinicians who are trying to reinvent the wheel and have not carefully read his research. They try to train frequency believing it has an advantage without actually doing research on it. This could theoretically just make people go to sleep or become very foggy headed. It is pushing the brain more than inviting the brain. There are also commensurate changes in other dimensions or EEG measurements known as phase, coherence, and symmetry that are involved that as well were not investigated at the time.

Green also experimented with different sounds, especially flute sounds like those in Adam Crane's Prism Five program that the clinical psychologist Nancy White used for AT training. One of his methods was to convert the lower brainwave frequencies into a band of harmonic frequencies that would be audible by multiplying their frequency by a factor of 200. He would have the frequency sound vary by tenths of Hertz which would result in almost musical tones. His idea was to provide the ability to hear the play of their brainwaves in an auditory form. That method has not been replicated to this day, but we plan on implementing it again in our software.

It should be apparent from the foregoing that Elmer experimented with a far greater variety of methods than anyone in contemporary times has tried. He did not, however, use two channels to train people although he had the ability to easily build as many amplifiers as he wanted. Presently, some people are using two channels to train what they call AT, but they have offered no research or formal explanation for it. Some practitioners have observed that the two-channel version helps those who have difficulty achieving crossovers but this effort to push the client may result in stronger abreactions. We have found that there is usually a good psychological reason that especially traumatized clients are resistant to crossovers and uncomfortable with the vulnerability they may experience in that state. This discomfort can be actually tracked using heart rate variability (HRV) monitoring or other biofeedback modalities.

Crossovers

Green noted that when monitoring these training sessions alpha tended to go up to 12 Hz and then gradually drop to 8 Hz. He observed trains of low frequency alpha and theta in the frequency band between 6 Hz and 8.5 Hz during periods of deep "reverie" as he called the liminal state he was trying to achieve. He further noted that the imagery occurs when the alpha suddenly drops in amplitude and long trains of theta appear after which a burst of "paradoxical alpha" occurs with the emergence of hypnagogic imagery. He comments that the key feature of the state are "hypnagogic and dream-like images during a state of semiconscious reverie…. in conjunction with periods of theta rhythm and low frequency alpha waves" (Green, 1977 p10). This he called a crossover.

This was the point at which subjects would have profound "creative insights" from their training that later they would process in a more active conscious state. The goal of the process was not just to achieve the liminal state, but to have creative insights. This is much the same process as meditation. The goal is not just the meditational state but the insight that arises from being in that state. It was this process that his psychiatrist friend referred to as dredging the unconscious. It should be clearly pointed out at this point however that this AT process is still a different process with a different goal than meditation pursues. The goal of meditation is not to generate creative solutions to problems. Physiologically the goal of meditation is to learn to enter an alert waking awareness in a state that is meant to convey a fundamental insight regarding the nature of reality and is marked by a consistent ongoing manifestation of low frequency alpha waves and perhaps other frequencies such a gamma or higher (Dunn et al., 1999; Austin, 1998). The goal is not to shift in and out of deep dream-like states in order to derive images. In order to achieve these deep alpha states, meditators practice for years in order to grow a brain that is capable of achieving these states (Davidson et al., 2003). Some of the confusion of this comes from the fact that alpha can drop into that frequency domain and remain morphologically alpha while theta is a morphologically distinct frequency in its own right. It is difficult to separate the two without an expert eye. Each frequency arises from a different region of the brain. Theta arises primarily from the subcortical septal hippocampal oscillators and the cortex through gamma coupling while alpha arises primarily from the resonance of the thalamocortical loops and the cortex (Meehan & Bressler, 2012; Kirk & Mackay, 2003). Theta is more associated with memory and emotional processing while thalamocortical loops are more associated with state shifts. Alpha theta appears to be a combination of state shift and memory and emotional stimulation while meditation is primarily a sustained state shift that aligns the frequency domains of the entire CNS in a unique pattern. More on this in the last chapter.

Peniston considered the crossover therapeutically significant because it usually resulted in an end of session report of a hypnogogic event or significant sensation, insight, etc. He felt the hypnogogic images were a critical component contributing to insight and integration (Peniston & Kulkosky, 1999). Peniston had also previously noted that vets received significant emotional relief from processing memories of combat in a highly relaxed state (Peniston & Kulkosky, 1991). White (1999) and Whuttke (1992) also saw crossover as an altered state associated with personal insight and integration. White, in particular, offered the theory that it was the character of the "witness state" that made it possible for clients to process the trauma without being retraumatized. Peniston & Kulkosky (1999) also theorized that subject recall of material was likely related to increases in the 13–26 Hz range during crossover (1999)—but recent research suggests that is more likely related increased muscle tension or bracing—which would be occurring during exposure and integration (Soutar, 2023).

Moore and Trudeau (1998) and Johnson and Bodenhamer-Davis (2017) both investigated crossover phenomena in more depth which has proven very helpful to the practitioner. Moore found that crossovers did not always produce hypnogogic images and yet clients still benefitted from AT training. They often had strong images in their mind's eye or just strong insights and key thoughts. Some never had any images and had instead strong physical experiences and sensory events such as floating, falling, flying, local tension or pain, etc. Moore also found that crossovers can result in colors and geometric patterns, biographical memories, symbolic images, and heaven forbid, spiritual experiences.

Johnson and Bodenhamer-Davis looked more at the crossover itself and used the same term of "liminal states" to describe them that we did, which is a common term in anthropological literature regarding states associated with primitive spiritual practices. Johnson and Bodenhamer-Davis (2017) believed that different combinations of bands resulted in different types of memory or sensory experiences—which makes sense if there are different styles of coping with trauma. They proposed alpha was related to free floating associations and mental imagery which is an observation going back to Barbara Brown's work in the 70s, and they did not realize high amplitude alpha was also identified as being related to anxiety and trauma (Kerson et al., 2009; Ros et al., 2013). That was observed by Scott (2017) however early on in his work with AT and when he trained alpha down and found that subjects felt relieved. Ros et al. (2013) also found training alpha down at Pz brought relief to individuals with PTSD. Part of the confusion regarding all this is that these authors are often focusing on amplitudes, and there are at least four other dimensions of neurometric analysis that also describe these states. It would be similar to describing different flowers with blurry black and white photographs. The neurometric dimensions of coherence, phase, symmetry, and dominant frequency are also critical in describing, defining, and distinguishing changes in states and cognitive functions that have high amplitude alpha yet generate different subjective experiences. Certainly, some degree of confirmation bias also plays a role in these traditional oversimplified attributions of component bands to specific states. One of the differences between a depressed person with high amplitude alpha and a meditating person with high amplitude alpha is the brainwave phase coherence (Banquet, 1973) as well as asymmetry (Choi et al., 2011).

Theta was supposed to be related to emotional memory, which it is based on modern neuroimaging, but it is also related to emotional valancing (LeDoux, 2015), insight, error detection (Wang et al., 2020), and cortical gamma production (Meehan & Bressler, 2012). It depends on whether you are looking at temporal, frontal, or posterior theta. Delta is related to limited external stimulus input—which it does but it can also be an indicator of stimulus importance with respect to the external attention system, inflammation, or sleep deprivation (Knyazev, 2012). This also depends on whether the delta is frontal global or local in nature in addition to the features of the other neurometric dimensions.

Again, citing old literature, Johnson and Bodenhamer-Davis argue beta increases dream awareness and elevated beta during crossover is likely an indicator of hypnogogic imagery emerging, and Bill Scott believed he saw the same thing. They argued from their data that a beta greater than 3.75 μV during crossover was a reliable indicator of a crossover image emerging. Unfortunately, the norm for this frequency range of beta for adults is in the 5 μV range (New Mind Normative Database N > 700) and an inspection of their data shows all of their subjects had unusually low power. This low power fast phenotype is common in alcoholics (Gunkelman & Cripe, 2008) but that is only possible if the beta is being filtered out of the EMG using advanced techniques such as independent components analysis that few researchers in this field even know how to perform or have equipment and software to access. Consequently, this is possibly very misleading information and needs to be further resolved.

Johnson and Bodenhamer-Davis wanted to define crossover type and quality in their research, which is admirable but there were several other questionable conclusions. One of their conclusions was that theta over alpha by a margin of 1 μV for 3–10 minutes with elevated beta resulted in optimal crossover. However, Scott observed that shorter crossovers, perhaps 2–3 minutes in our estimation, produce more images than longer ones, citing yet unpublished research at Harvard by Lowen with and n = 250 (Johnson & Bodenhamer-Davis only has an n = 10). This is more consistent with our clinical observations as longer crossovers, as Scott also observes, usually involve deeper states that often are closer to sleep.

Johnson and Bodenhamer-Davis used a classification system and statistical analysis to rate crossover experiences and tie them to frequency band combinations. Experiences fell into the following categories in their opinion:

Abstract—Sensory and emotional experiences

Biographical—What happened and forgotten

Perinatal—Birth experiences—rebirth experiences

Transpersonal—Transcendental experiences, symbolic and archetypal

Based on his research the occurrence of these categories was as follows: 27.9% had no crossovers, 34.1% had spontaneous imagery or memory events, 37.9% spontaneous sensory experiences related to memory, biographic images occurred 7.1% of the time, perinatal images 2.1%, and transpersonal 6%. They fortunately used a rigorous crosstab analysis for these conclusions, which should be helpful as a reference for clinicians. In his group of subjects, during crossover, clients had therapeutical useful imagery occurred 85.7% of the time—which indicates what a valuable opportunity it is to review the images and use clinical skills to analyze them.

Johnson and Bodenhamer-Davis theorized that delta was related to transpersonal experience and consequently it was likely not wise to inhibit it. Scott admits that his discomfort with intense emotion abreactions and concern for subjects dropping out of his research lead him to avoid abreactions. If fact he did not believe that clients had to be uncomfortable to heal (Scott, personal communication). Johnson, like Peniston, believed intense emotion was indeed a part of the healing experience and we agree. Research in trauma is also supportive of this conclusion (Forbes et al., 2020). Suppressing delta may be suppressing transpersonal and spiritual aspects of integration (Kakumanu et al.,

2018) but there is presently only a newly emerging literature on the topic although one renowned yogi and spiritual philosopher, Ken Wilbur, recorded global elevated delta while entering what he claims is Nirvakalpa Samadhi (Wilbur, 2000).

Bill Scott also suppressed lower frequency theta as he set the theta filters for 5–7 Hz instead of 4–7 Hz to avoid abreactions. This was later supported by the finding that low frequency theta tends to carry a more emotional component of memory information and is processed through the amygdalofugal pathway rather than higher frequency theta which is processed through the mammillothalamic pathway (Kirk & MacKay, 2003). In contrast, like Peniston and others, we found that abreaction was a critical component of the healing process and preferred the 4–7 Hz setting for theta. In addition, they found from their statistical analysis that higher amplitudes of slower waves usually indicated they would progress from biographical to transpersonal. This should also be a helpful clinical indicator of where to place the clinical emphasis as clients progress and who is likely to be more receptive to a more transpersonal therapeutic approach. Another observation we have made was also confirmed by this valuable research project. The Beck, Pittsburg Sleep Scale and MMPI were used to determine if therapeutic outcomes were related to frequency of crossovers, and this was in fact confirmed. These were also the subjects, it turns out, with most transpersonal experiences.

With respect to filter settings, Johnson and Bodenhamer-Davis set their alpha threshold parameters for a 50–70% reinforcement rate and their theta thresholds for 20–30% reinforcement rate. Since they tried to follow Peniston as closely as possible, they chose to train at the sensor location O1 to follow. This is in contrast to Scott who prefers Pz and has had excellent outcomes with larger subject samples. This is still an issue of disagreement in the field that is not likely to be resolved without more robust research on the topic. Either location appears to produce good results.

Johnson and Bodenhamer-Davis preferred a dual state of consciousness model of integration that focuses on the importance of clients witnessing past traumatic events in calm state. This agrees with the most recent therapeutic approaches to complex trauma (Forbes et al., 2020) as well as the observations of van der Kolk (2014) in his decades of work with complex trauma. Nancy White (1999) also subscribed to this perspective. We at New Mind also appreciated Johnson and Bodenhamer-Davis's arguments that clinicians should be versed in symbolic interpretation because a lot of it occurs in AT and it can be of crucial importance in the clients healing process. Many of us rely on the works of Joseph Campbell and Carl Jung for resources in this area. Martin Wuttke (1992), in the early years of AT method development, also noted the important role of transpersonal imagery in integration process. These investigators argued that time between sessions for clients to digest experiences was also an important consideration and integrations could not be rushed. I hear this sentiment as well from most of the practitioners I have consulted with or mentored over the years. In addition, these investigators also recommended fast wave training based on the qEEG to correct for issues arising from drug abuse etc. which is highly comorbid with PTSD. The fast wave training improves inhibitory control as does HRV or biofeedback training and this is an important component of stabilization in the ISTSS model of complex trauma therapy.

The Peniston Procedures

Peniston took a workshop from Green, as was mentioned previously, and decided to employ his methods with alcoholics in what is now, many decades later, a famous experiment. Unfortunately, the methods section of his article was not clearly written and for years there was confusion and arguments over what Peniston did, as well as many resulting innovations, until he published a chapter in Quantitative EEG and Neurofeedback (Peniston & Kulkosky, 1999). The following is taken from that explanation.

Peniston had found that EMG biofeedback, that is, teaching clients to reduce muscle tension, reduced depression scores on the MMPI (Peniston & Kulkosky, 1989) and he observed it was closely associated with alcoholism. He felt that EEG related relaxation techniques would also assist based on the work of Kamiya, Brown, and Green. To make a further impact he decided to combine these various methods with other methods being used at the time including systematic desensitization, temperature (TMP) biofeedback, visualization training, rhythmic breathing, autogenic training, guided imagery and constructed visualization. Hence the multi-modal approach.

Patients received an EEG assessment upon admission to the hospital and were evaluated for alpha amplitude, dominant frequency, percent power and phase synchrony. They were given Beck Inventories, MMPIs and a Catell 16PF Inventory. Autogenic training exercises were used with temperature training and rhythmic breathing techniques for a period of 30 minutes as pre-training and each patient received 6–7 sessions. These sessions were adapted from Green's research (Green & Green, 1977). The temperature training was thought to stimulate theta and a reverie state. The visualization training involved seeing themselves physiologically relaxed, as having the personality they would like, and then seeing themselves go through their daily life acting out the qualities of this type of personality.

The sequence

A 5-minute eyes closed baseline threshold is set for theta and alpha. A 5-minute eyes open baseline threshold is set for beta. The patient then practices his visualization with eyes closed and is instructed to sink down into the theta state and then instructed to proceed with the command "Do it." The volume of theta and alpha are independently set to "comfortable levels" while beta threshold volume is turned off and the patient is left alone in the room for 30 minutes. When the training session comes to an end, the patient then completes a questionnaire. This process is repeated with each patient for 28 consecutive days.

Abreactions

Peniston clearly states that abreactions could occur, especially if theta reinforcement was set too high. He recommends making adjustments over the following six sessions to reduce the theta. They even had a procedure for dealing with it. The beta tone was turned on and theta threshold was increased so there was less reinforcement of theta and alpha tone turned off. The client was returned to the reverie state for 25 minutes to recover and calm down. The patient was trained like this for the next six sessions. Over time this was simplified by practitioners who followed to letting the client complete the session and training SMR 10 to 20 minutes at the end of the session. So, there was an evolution and variations that emerged based on changes in the technology and further extensive experience. Peniston does not offer any extensive rationale and much of it seems contradictory or counterintuitive. Unfortunately, that was typical of his documentation of his procedures.

Settings

Peniston only uptrained his patients and did not use inhibits like Bill Scott. He set beta at 40–50% reinforcement (this particular beta setting was done over a 5-minute baseline with eyes open), alpha at 50% reinforcement, theta at 20–30% reinforcement and ignored delta. Interestingly, these settings are considered too low for reinforcement by today's standards. I am assuming that the beta reinforcement at 50% was used to inform patients they were too tense and so it acted as a form of inhibit because that was the strategy off Bill Scott. Peniston is not clear on this issue, like many others. The beta filter frequency range was set at 13–16 Hz, alpha was set at 8–13 Hz, and theta was set at 4–8 Hz. The theta filters in this case would be picking up low frequency alpha waves and uptraining them along with the theta. These are very different settings than used by Green and apparently of Peniston's own invention. Hopefully the reader can appreciate at this point that this is a little more involved than just slapping electrodes on someone's head and running a canned protocol.

Nancy White's Theories

Nancy White wrote a chapter exploring the theory and rationale behind neurofeedback that was extremely cogent and a significant contribution to AT theory (White, 1999). Nancy proposes that an important aspect of the training is programming the unconscious with mental rehearsals of new images and intentions. She also cites the literature on the use of imagination in healing that was used by traditional shamans and contemporary psychotherapists of her time such as Dean Ornish, Norman Shealy, Larry Dossey, and others. She identified the procedure as falling into the category of transpersonal psychology and utilizes Maslow's paradigm to help explain its mechanisms. Theta waves are identified as internal processes and beta waves as external processes with alpha as a bridge. In this she and others anticipate the self-referential aspect of the default mode network and the efferent fiber pathways from the septa-hippocampal oscillator circuit that carry theta frequencies to this hub as well as the central executive network and the cortical activity it generates in beta in the frontal regions (Barnett et al., 2019; Buzsaki, 2006; Mitchell et al., 2008). This was, by the way, a fairly conventional perspective held by many involved in this research at the time, but there were few workshops on the topic.

Anna Wise (1997) developed a group therapy approach around this perspective and utilized EEG monitoring as part of her methods in workshops at Esalen Institute. Instead of AT biofeedback, Ann used visualizations and exercises while monitoring the changes in her clients' brainwaves. Disorders resulting from physical and emotional traumas can in many instances result in low alpha amplitudes that prevent one from accessing internal theta states. Wise also saw this as a block to integration and healing.

For Nancy White, training in alpha amplitudes seemed logically to increase internal connectivity. She proposes that the liminal states of consciousness at the "edge" of waking and sleeping are corridors to the unconscious. This is of course a very old idea in psychotherapy and was part of hypnosis theory. Theta is also associated with childhood because it is the dominant frequency during that developmental period. She reasoned that it might provide for state dependent memories to emerge.

Tom Budzynski had promoted the term "rescripting" for theta state learning and conducted research in this area (Budzynski, 1999). There is the opportunity to "reassociate and reorganize" in these states or change the basic templates for behavior that arose during development (Rossi, 1986). Survival patterns of childhood, Budzynski and Rossi reasoned, may become the foundation for future maladaptive behaviors and rescripting could be an effective means of updating and changing these behaviors. As a consequence of the ACE study, we presently now have firm evidence that that is indeed the case and the term toxic stress has been coined to describe it.

Nancy also proposes a neurochemical model to partially explain the effects of AT, but these are unnecessary and do not stand up in light of more recent findings in neuroimaging. The neuromodulator theory of depression has been determined to be incorrect as well as many of the other related theories (Cowen & Browning, 2015). She also cites the work of Bruce Perry promoting the concept that severe emotional trauma in childhood results in altered development of the CNS and that such changes could bias the system toward PTSD and other disorders when stressed in adulthood. Again, today we have the evidence to prove this theory in no uncertain terms as explained by van der Kolk (2014). She favors the idea that changing brainwaves can change neurochemistry and we now know that because excitotoxicity from anxiety engenders excessive glutamate increasing inflammation. At the same time reducing arousal enhances conversion of glutamate to glutamine which parallels the reduction of inflammation (Wang & Michaelis 2010). In her theory she also explores the idea that the therapist's presence is a crucial feature of the process, and this too has received broad support in contemporary research (Forbes et al., 2020).

Building on theories that all people have core traumas and that these guide future behavior patterns is consistent with the entire theoretical body of knowledge in clinical psychology going all the way back to Freud. Freud called complex trauma neurosis and saw it as a fundamental aspect of ego development. Reviewing the work of Stanislav Grof (1993), she proposes that flashbacks related to trauma can emerge in the form of hypnogogic imagery. For her, Grof's COEX system explains memories as thematic systems bound by emotion that are chronologically layered which is still an interesting and plausible theory. Accessing these results is catharsis that discharges the emotional distress associated with the memory. Although she does not specifically explore this mechanism, Joseph LeDoux (1996; 2002) provides a detailed biochemical and neuroanatomical model to explain and support this perspective. Finally, she introduces the transpersonal psychology theories and research that begin to explain the role of consciousness in this process of recovery and integration and connects it with more traditional modes of healing found in shamanism. White cites the work of Bohm (1995) and Goswami (1993) as providing various models integrating consciousness and physics and my own research echoes her findings. There is presently a long list of others who have followed with similar perspectives. Additionally, she discusses the value of Campbell's (1988) idea of the Hero's Journey as a fundamental motif that each person expresses in a different manner in their life quest to integrate the challenge of life experience. All of this we will explore deeper in the final chapter as this is the aspect of Elmer's theories that were overlooked by skeptics because they were not fashionably scientific. Finally, White proposes that the theta state allows the experience of trauma from a nonjudgmental transcendent perspective that results in integration of the traumatic experience, and this too is consistent with van der Kolk's conclusions although he might avoid the transcendental components of her arguments. Overall, this is still a very strong integrated model for AT training that is consistent with our own understanding at New Mind.

Other Theories

Adam Crane's (Crane & Soutar, 1999) theories of negative rehearsal and trances derived from Stephen Wolinsky (1991) fall along the same lines as White. My own theories outlined in the Automatic Self (Soutar, 2015) include all of the foregoing and are derived from the integration of a large number of theoretical paradigms as well. Those engaged in this "process," as Crane referred to the alpha theta experience, often emerge with very similar perspectives from their clinical experience. Adam and I explored these similar ideas that we also discovered in our parallel experiences with AT training. So, there appears to be considerable independent agreement emerging from different experiences in time with this process of AT. I will expand and refine these ideas later on in the book.

Bill Scott has never formally outlined a theory of AT, but his methods and workshops reflect a perspective very consistent with those outlined in the works of the previous authors cited. Bill's contribution has been the refinement of technique through research. As we have documented, Bill has done several studies which have been very valuable in illuminating this method. His work with Native Americans in collaboration with Peniston resulted in many other important insights about the method of AT. As we have shown, his work at UCLA department of Psychiatry resulting in the CRI-Help study (Scott et al., 2005) mentioned previously demonstrated that NFB could normalize TOVAs in 14 days when done twice a day. No one reported and documented that NFB could generate changes that quickly until this was published. It also showed the power of AT to deal with multiple addiction patients in institutionalized settings, replicating Peniston's work. In this process, by specifically including 12-step methods in the research he was able to contrast it with Ken Graap's findings that AT was not especially effective with addictions without the 12-step component (Graap & Freides, 1998; Scott et al., 2017). In addition, Bill was the first to report the dynamics of crossovers and how they related to hypnogogic imagery, which we will cover in the section on crossovers. These technical findings did a great deal to expand our techniques and efficacy in AT training.

Chapter 4

Contemporary Trauma Theory and AT Psychophysiology

Trauma and Addictions Theory

The various forms of trauma have many features in common. As mentioned in the first chapter, the more obvious features and their consequences have been well understood since the early 20th century. The more subtle features have been slowly percolating into our collective therapeutic consciousness slowly since the 1960s and 1970s. For many of us, the subtle features of complex trauma became obvious in the field of addictions and family systems, but the concept of PTSD was still novel at the time and we had none of the present terminology to describe it in the manner that we presently do. For me, Bradshaw was the leading theorist and publicist for many of the basic concepts but there were many others who did fundamental work in this area. In my master's thesis on codependency and the family system I reviewed all of this literature in detail, and it is still a good summation of what we knew at the time. The Bowen family systems theory and the work of Virginia Satire and others at Mt. Sinai hospital was groundbreaking. But before that we had Fritz Perls and Ida Rolf at Esalen Institute in California pioneering work on the psychodynamic and physiological features of trauma.

Looking back, it is easy to say that complex trauma clearly fueled the transgenerational behavior that drove addiction. When I proposed that idea to my graduate advisor in the 1980s, he agreed but we both knew the academic community was not yet ready for that concept. It has taken decades to be recognized and accepted and much of it has been driven by the research into the biology of trauma. In parallel, the literature on TBI in the 90s was equally struggling alongside the then heretical concept of neuroplasticity to be recognized as a valid conceptual approach to trauma. Hard as it may be to believe today, the idea that TBI could generate anxiety and panic disorders as well as depression due to neuronal shearing was anathema. This was the academic and policy driven environment of the time when neurofeedback emerged as a technology. As mentioned in the previous chapter much of this repudiation and denial was driven by fundamental cultural bias grounded in 19th century scientific positivism and the conception of man as a primarily rational actor. In both economics and psychology that concept has been put to bed; ironic considering the irrationality of the world wars after the Enlightenment. The demise of the rational actor in economics did not even mature until recently with the Nobel-winning work of Daniel Kahneman in economics as did the brilliant work in the 90s of Daniel Goleman on emotional intelligence. For me, Walter Freeman put the last nail in the coffin in his 2009 paper on qEEG and fMRI declaring that it was clear from the neuroimaging research that man was primarily an emotional decision maker, echoing the philosophical observation of Aristotle.

The literature on codependency identified key features of complex emotional trauma arising from addictive behavior and the related abuse that it generated. It identified the family secret as the culture agent of denial that kept everyone from looking closely at all of the dirty little secrets. As van der Kolk remarked, it was strange that the textbooks in the 60s declared that incest incidence rate for women was only one in 1 million and yet they all showed up at his clinic. Now we know the truth. Cultural denial was as broad then as in the Victorian era when Freud struggled a similar observation. Now we know that conservatively it is 10% for incest and one in five women are sexually abused by the age of 19. Men have significant abuse demographics as well. Alcoholism, drug addiction, gambling addiction, and food addiction all have similar consequences and arise out of complex trauma with the right genetic polymorphisms. The diathesis stress model of mental disorder seems to me stronger today than when I learned about it in abnormal psychology in the 80s. A combination of genetics and cultural and familial stress combine in toxic patterns of transgenerational behavior that propagate the maladaptive and destructive behaviors that we now identify as complex trauma.

Fundamental Features

I will first discuss the features of acute and chronic PTSD both in terms of behavioral symptoms, but I will also explore trauma in terms of electrophysiology. The maps patterns we consistently see in the clinics are often a good indicator of what is going on. The most salient features of PTSD are re-experiencing, avoidance, emotional numbing, and hyperarousal.

Re-experiencing involves recurrent episodes of flashbacks, nightmares, repetitive distressing images and/or sensations. There is often a focus on thoughts about how the client could have prevented the trauma that involve feelings of guilt and shame. This process also involves stimulus events around them that become triggers for the flashbacks. These events often involve bursts of theta frequencies that carry the memories from subcortical sources. The memories are often fragmented and can be experienced as separate sensory events. People that have frequent dissociative episodes like this often have very high theta waves in their EEG (as shown below) that come and go in amplitude in relation to their intensity. These waves associated with dream-like states, imaginative episodes and intrusive images. These people are usually absorbed in their DMN self-referential system and act distracted.

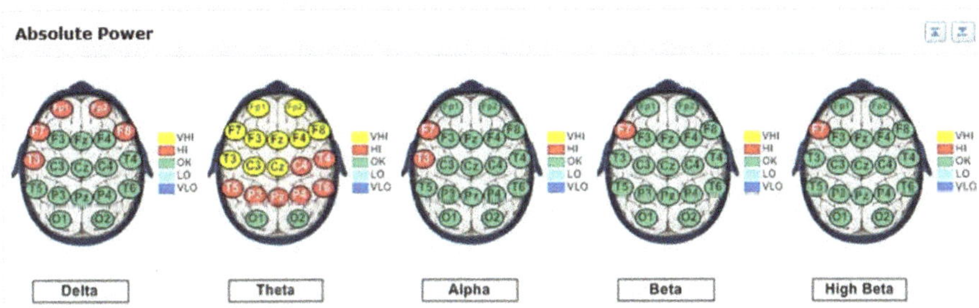

Avoidance involves avoiding certain people, places and things that can trigger memories. It also includes avoiding discussion about the traumatizing experience and the related emotional and physiological side effects. A key method for this may be distractions such as hobbies or workaholism and often includes considerable "mind chatter" that manifests as elevated alpha, especially in the posterior DMN region (as shown below). Interestingly, in contrast, Ros et al (2013) have observed this mind chatter in low alpha maps as well. These individuals find that their mind is always going, and they habituate to it to the point where they are generally not aware how omnipresent the chatter is until they try to go to sleep. They commonly present with sleep onset insomnia. Intrusion was eventually added into the DSM-5 symptom list and typically manifests in this EEG subtype as well.

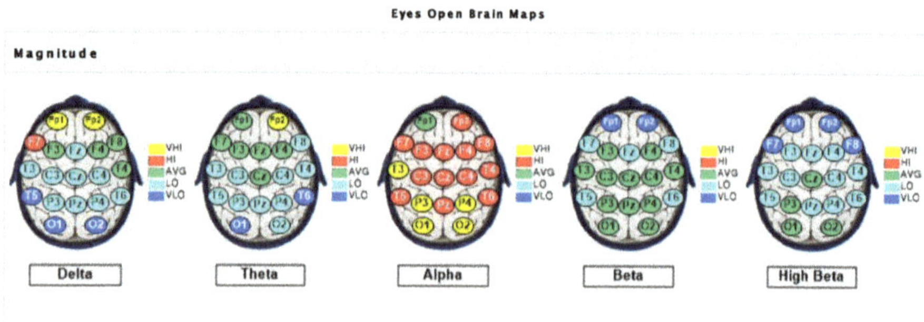

Another is emotional numbing or blocking the traumatic emotional memory pathways which often presents as very low levels of theta frequencies and theta hypocoherence, especially in the temporal regions involving emotional valancing. In the example below theta connectivity is low in the temporal region and theta is very low in the frontal region blocking the frontal pathways from the anterior septal hippocampal pathway.

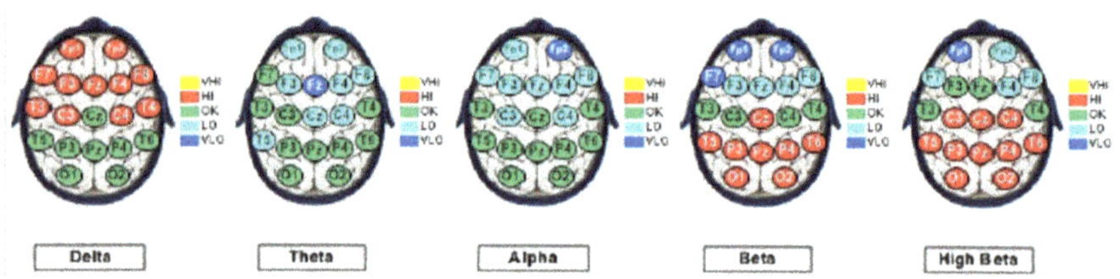

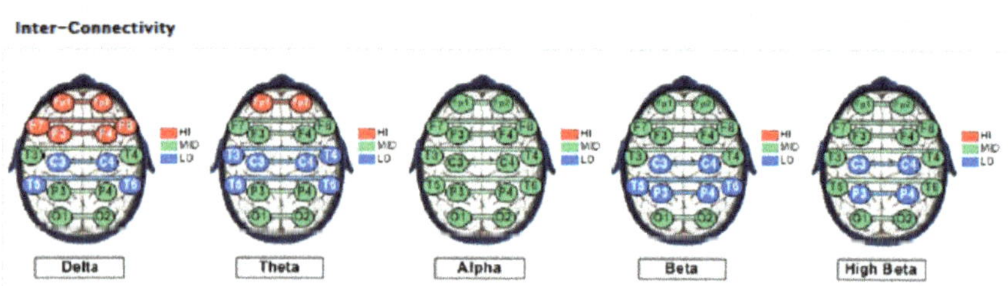

Hyperarousal and associated hypervigilance is another feature which may also present as an exaggerated startle response and hypersensitivity to environmental stimuli. This often involves high levels of anxiety involving irritability, angry outbursts, insomnia, and difficulty concentrating. The concentration issues are usually a consequence of anxiety and poor sleep which undermines hippocampal function. We see low posterior theta indicating the posterior septal hippocampal pathway to the DMN is diminished in function reducing emotional inputs to this self-referencing center. The angry outbursts often are driven by surges in arousal that shift beta frequencies shifting to the right hemisphere as norepinephrine spikes and clients feel out of control emotionally as they perceive they are losing control of their environment. Often there are additional posterior coherence issues in the beta frequencies indicating poor sensory processing and excess sensitivity to light.

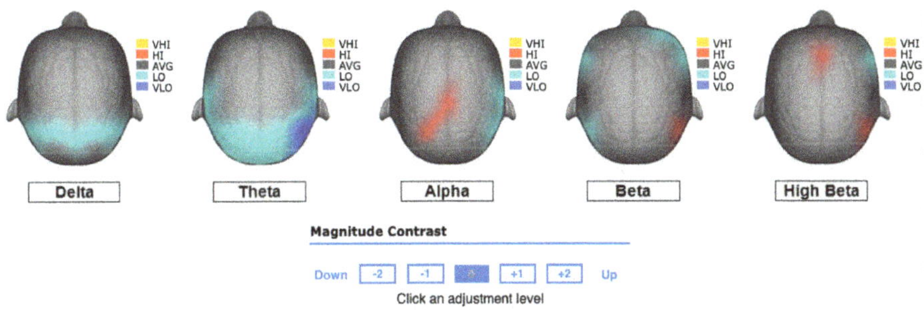

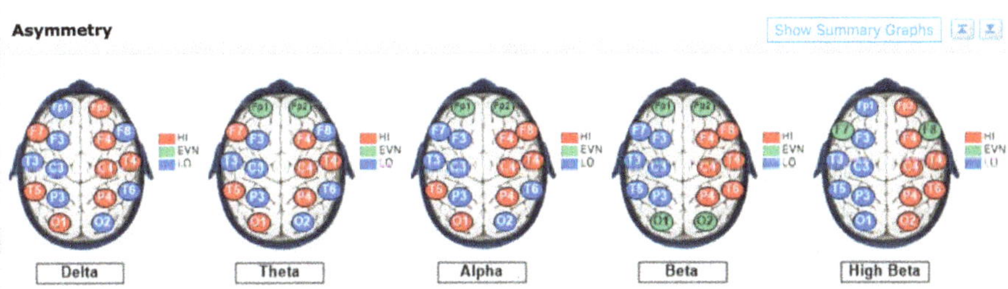

Most people experiencing chronic hyperarousal and hypervigilance who somaticize their trauma express it in chronic muscle contraction known in psychophysiology as bracing. Most frequently, this can be readily observed in the scalp muscles and manifests on the brainmap as high amplitude hi beta. This is known as artifact because it is not real EEG but instead muscle activity and this EMG muscle artifact contaminates beta frequencies as well. Clients with this feature will become progressively exhausted easily as discussed in the physiology section and become depressed.

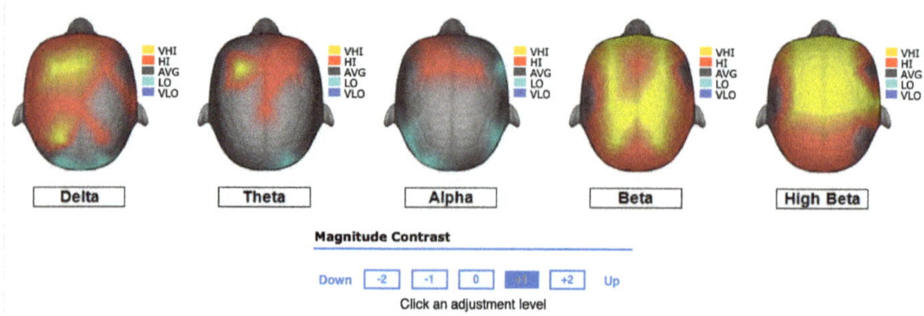

ISTSS Guidelines (https://istss.org/home) suggest that guideline development is an imprecise science and varies greatly with clinical intervention methods. Methods should consider both randomized controlled trials (RCT) as well as clinical experience balancing both against client needs and presentation. EMDR was considered highly recommended, but NFB was never reviewed. Our clinical experience over the last 30 years was that AT was typically more effective at the clinical level when deploying both methods in the same clinic. Their book based on the study generating these guidelines, Effective Treatments for PTSD (Forbes et al., 2020), however, does mention neurofeedback as a promising approach, citing two RCT studies demonstrating efficacy under the rubric of "Other Biological Treatments."

Treatments are usually focused on the trauma or on non-trauma solutions such as relaxation, stress reduction, and interpersonal therapy (Watkins et al., 2018). Emotional processing theory suggests that traumatic events are not emotionally processed at the time of the traumatic event and remain as a cognitive structure which includes feared stimuli, feared responses, and associated meanings. To alter these fear structures, they must be activated at the same time new information is provided that is incompatible with the fearful meaning of the structure. In vivo or imaginal events are invoked and breathing training is involved. The individual repeatedly recounts the narrative in the present until new adaptive information is integrated. This is a highly effective approach compared to supportive counseling and non-trauma-based therapies. Across studies 41–95% resolved their diagnosis (PTSD not complex trauma).

Cognitive processing therapy (CPT): Initial sessions involve psychoeducation regarding the cognitive model of trauma and an exploration of the client's conceptualization of the traumatic event. Why it occurred, how it has changed beliefs and self-concepts and how it affects safety, intimacy, trust, power and control, as well as self-esteem. It also reviews how their view of the world and others has been changed by the trauma. Connections between thoughts, feelings, emotions and the client's stuck points are explored. Clients are encouraged to challenge maladaptive beliefs. CPT has been shown to be very effective at reducing symptoms of PTSD. Integration of trauma and loss of diagnosis ranged from 30% to 90% in published studies.

Cognitive behavioral therapy (CBT): The APA encourages use of trauma-focused CBT and non-trauma focused CBT. Individuals with PTSD engage in problematic behaviors and cognitive strategies that prevent them from changing negative appraisals and trauma memories. Correcting autobiographical memory, removing maladaptive behavioral and cognitive strategies and managing negative appraisals are key remedies. Guilt appraisals that are paired with negative cognitive activity are explored and resolved. Exposure and cognitive restructuring are key methods. Imaginal exposure can involve writing and reading the trauma narrative out loud. Focusing on management of in vivo exposure is also used. Cognitive restructuring involves identifying dysfunctional thoughts, eliciting alternative rational thoughts, and reappraising beliefs about self, trauma event, and the world. Loss of diagnosis is 61–82.4%. Far superior to relaxation. Aggregate dropout rate across all three methods was 18.28% (so similar rates are likely to occur with NFB).

Although effective for acute PTSD, these methods may not be as effective with complex trauma. The reason for this is that many clients may not be aware they have CPTSD or have any memory of events that may have precipitated it. In addition, there may be ongoing multiple exposures that require repeated imaginal exposures to work through. The good news is that these exposure-based therapies do not have any reported increase in adverse side effects when compared to other therapies. Military samples however showed that 60–72% of veteran population continued to retain their diagnosis after treatment. Van der Kolk (2014) noted this difference in military trauma involving veterans as well. This population obviously represents aspects of trauma still unaccounted for in the literature and additional features we need to consider, both neurological and sociological.

Complex Trauma and Toxic Stress

The symptoms of complex PTSD include those mentioned above but also include feelings of guilt or shame, difficulty controlling emotions, and periods of dissociation. Physical symptoms may include headaches, dizziness, chest pains and stomach aches as well as many others.
The fact is, this list is not realistic. When looking at the literature in psychology on physical symptoms of anxiety, the items number over 70 at minimum. At any rate, traumatized individuals will commonly reduce their interaction with others to the point of cutting off friends and family. It should be pointed out at this point that this is a common feature of chronic anxiety and as well as its fully mature manifestation of depression. This avoidance usually arises because of the destructive relationship behaviors arising from the other symptoms makes interaction painful and counterproductive. Some engage in unusual risky behavior that may involve self-harm, directly or indirectly, and alcohol and drug abuse. Suicidal thoughts are often present because of the growing sense of shame derived from social interaction failure.

The International Society for Traumatic Stress Studies (ISTSS) has officially published criteria for diagnosis and recommendations for treatment. This is partially in response to the DSM-5 position that it should just be subsumed under the category of PTSD. They argue that CPTSD is a more complex disorder requiring a more extensive course of treatment than PTSD.

The identified sources are:

Inability to escape repeated trauma

Attachment trauma—child abuse or domestic abuse

Ongoing armed conflict—prisoner of war

Human trafficking

Chronic illness

Strict religious traditions or cults

Symptoms are as follows:

Prominent dissociation

Difficulties with attention

Difficulty in regulation of affective impulses managing anger and self-harm.

Somatization and or medical problems

Altered perception of perpetrator

Dysfunctional avoidance of relationships—no trust or intimacy; others are out to exploit them

Altered self-perception—internalization of abuse messages, worthlessness, guilt, shame

Altered systems of meaning—sense of hopelessness for help.

The ISTSS promotes a three-phase approach to treatment. This approach is very similar to what was evolved in our neurofeedback clinics over a period of two decades. This was in part because our model of trauma grew out of the addictions and family systems model of trauma which in retrospect was a very refined theoretical model with a psycho-social perspective. Much of this as mentioned elsewhere is because of the pioneering work of Karen Black and John Bradshaw as well as many others. Our model was combined with the initial research done at the Menninger Clinic in Topeka, Kansas, which we will continue to discuss, which also included physiological measures such as beta endorphin levels as well as methods involving biofeedback and breathing exercises. In addition, group meetings added a social component to the Menninger protocol and now we know the meetings were actually 12-step meetings. This constituted an early bio-psycho-social approach to treatment with many of the components of tradition AA perspectives and methods.

The ISTSS phase-based approach involves:

1. Safety stabilization and emotional skills building

2. Processing memories of traumatic experiences

3. Transition into greater engagement and everyday life

In our experience phases 1 and 2 are typically recognized to take 9–12 months in the form of the initial NFB fast wave training followed by AT training, while phase 3 usually involves another 6 months.

Phase 1 seeks to establish patient safety to reduce potential retraumatization. This may involve helping individuals remove themselves from the environment generating or sustaining the trauma.

Developing a therapeutic alliance building on mutual trust is critical to this phase. This also helps provide the client confidence in their ability to learn how to build a successful intimate relationship and for the therapist to model its dynamics. It allows clients to address relationship patterns and how to change them through development of agency and healthy boundary setting as well as flexible responses. In NFB work this change often comes spontaneously as a consequence of enhanced self-awareness and reduced reactivity in conjunction with increased creativity.
Clients are also educated about the nature of trauma so they can better understand their behavior and the course of treatment. They learn that CPTSD is something they have and not something they are. This insight is also something that naturally arises from AT training.

The emotional skill component helps clients identify how they have disconnected from overwhelming emotions and are shutting down. They learn emotional self-awareness techniques and how to manage emotions through centering and breathing. Mindfulness training may also be a part of this stage. This is similar to using standard NFB and biofeedback to calm the system and enhance sleep with the addition of HRV training to develop better breath control and arousal management. Part of this effort also involves CBT therapy which as noted above is often quite effective with PTSD.

Phase 2 utilizes exposure therapy in a safe environment by reviewing the experiences of the trauma that generates less reactivity over time in part using the skills built in Phase 1 to manage, process, and integrate the emotional components of the trauma. For more extreme cases, suicidality may emerge from this process, in which case a return to Phase 1 is crucial for further integration. Changing the clients' beliefs about the past helps them become unstuck and aids in building a new self-concept. This phase is similar to AT training where clients naturally begin to remember and integrate trauma experience in accordance with and in harmony with their physiological and psychological sense of safety.

Phase 3 is a stage where clients consolidate their gains through building a new self-narrative involving future plans, enhancing relationships, developing safe employment, pursing new education goals as well as self-nurturing recreation and hobbies. This phase reflects the physiologically based findings and theories of Joseph LeDoux with regard to the natural pattern of consolidating implicit memories and moving them into explicit memories comprising a component of the individual's life narrative and identity. This is often a period where we found people tended to move in a spiritual direction to find meaning in the trauma that contributed to their personal growth at the deepest level. This component moves beyond the scope of practice of most therapists, if not their comfort zone, and consequently is not adequately addressed in most models of trauma. This was however, recognized as a major issue in the power struggles in the early AA movement as they labored over whether to include the term God or the term "a higher power" into the AA 12-step methodology.

As the reader will see, all of these components and more are included in the NewMind AT approach but based more on a psychophysiological model grounded in research. This will be discussed in detail in the AT methods chapters.

Toxic Stress: A Medical Paradigm

When the features of complex trauma arise from childhood experiences, it is often referred to as developmental trauma, while development of chronic PTSD in adulthood usually bears the label of complex trauma. It should be noted that developmental trauma is associated with the term toxic stress in the medical literature, especially the pediatric literature. Shonkoff and Garner (2012) wrote the position paper for the American Academy of Pediatrics on this topic and it is based on the Adverse Childhood Experiences (ACE) study.

According to the CDC, the Adverse Childhood Experiences study was published in 1998 and was a collaboration between the Centers for Disease Control and Kaiser Permanente. It assessed the relationship between adult health issues and childhood abuse as well as household dysfunction. The study began with 9,508 individuals with a follow up study sample involving 17,337 individuals. The study clearly showed that adverse childhood events resulted in an increased risk for smoking, alcohol and drug abuse. It also showed increased risk for depression, suicide, poor health, large numbers of sexual partners, and challenges with physical inactivity and obesity. (Please recall that approximately 40% of the American public is obese.) Individuals were also more likely to attempt suicide, have increased risk for broken bones, heart disease, lung cancer, liver disease and multiple types of cancer. Impaired job functioning was common in this population as was the risk of homelessness and criminal justice involvement. Obviously a technology that has a significant impact on this population would be revolutionary.

Shonkoff and Garner (2012) argue that the findings suggest that many adult diseases should be viewed as developmental disorders. Most debates around childhood poverty focus on educational objectives rather than considering the impact of childhood adversity and its long-term consequences. Stress has been characterized in this medical model as positive, tolerable, or toxic. Positive stress involved the presence of parental emotional support that helps the child cope and return to baseline arousal when challenging stressors occur. Tolerable stress is defined as stress derived from nonnormative experiences that carry greater adversity such as death of family member, serious illness or injury or a contentious divorce. In a supportive environment, the consequences of such events is greatly diminished from a psychophysiological perspective. Adaptive coping and a sense of control are critical components for the child to manage stressful events. Toxic stress involves strong, frequent and prolonged activation of the physical stress response. This can involve child abuse, neglect, substance abuse, and mental disorders. By defining trauma in terms of ability to adapt to circumstances despite stress, this paradigm has provided a link to the general adaptation model of stress and the extensive research on stress in psychology in the latter half of the twentieth century. This is also the same model used by Herbert Benson in his research at Harvard and past publications on the relaxation response (Benson, 2000) as well as his more contemporary research. His research demonstrates that we have innate skills for dealing with stress and once learned they have a profound impact on the cumulative physiological as well as psychological effects of stress, even at the level of DNA and RNA. The flaw in his proposed techniques is not the techniques themselves but implementation at the public level. This where biofeedback excels because we are a computer and media-oriented society and this is where services have to be delivered in an entertaining and palatable fashion.

Since biofeedback arises from the field of psychophysiology and uses stress as a primary dimension of concern, this approach to trauma also invites the cross disciplinary bridges to link neurofeedback and biofeedback both to research and development in this area of investigation. The section on physiological and neurological consequences of stress investigates these consequences in more detail. In particular the ACE model outlines how toxic stress changes the structure and function of the body and the brain and Benson's research lays the groundwork for the solution along with the work of Green & Green (1977) and other major investigators in the field of psychophysiology. Mental disorder is not a dynamic and dysfunctional state of the brain and mind alone but it is based on structural and functional changes due to prolonged stress response. We grow an anxious or depressed brain.

The long-term consequences of toxic stress are profound and affect every aspect of the organism as follows.

Health effect of toxic stress:

- Diminished immune function
- Inflammatory process
- Neural excitotoxicity
- Telomere shortening—early aging
- Abnormal fat deposits
- Arterial degradation

How stress affects structure and function of the brain:

- Hypertrophy of amygdala—hypersensitivity to stress
- Hypertrophy of orbital frontal cortex—reduced socio-emotional contextual evaluation
- Reduced hippocampal function—degradation of memory function
- Reduced medial PFC function—loss of top-down control and increased impulsivity
- Neuronal death—loss of cognitive resources
- Degradation of neuronal connections—loss of cognitive function

Toxic stress degrades cognitive abilities:

- General cognitive
- Linguistic abilities
- Socio-emotional skills
- Developmental landmarks

- Decision-making
- Working and short term memory
- Behavioral self-regulation
- Mood and impulse control

These findings demonstrate that toxic stress impacts all aspects of physiological function and especially the dimensions of psych-social and social-emotional function. This in turn impacts all aspects of social relations and interactions.

Anxiety, Depression, and Trauma

It is my thesis that anxiety and depression are inevitable parts of all mental disorder. This is well reflected in the literature. Excitotoxicity from emotional chronic hyperarousal results in rapidly developing neuroinflammation along with global elevations of adrenaline, cortisol, and glucocorticoids due to hypothalamic and amygdalic upregulating of the hypothalamic pituitary axis (HPA) and the central nervous system (CNS). This is discussed in detail in the physiology section next chapter. We refer to this chronically elevated psychophysiological state and level of excitotoxic action in psychology as anxiety and it has well documented global physiological ramifications.

Depression emerges from the exhausted state arising from chronic hyperarousal of anxiety and the associated sleep degradation and deprivation. It is part of a bi-directional inflammatory response to stress occurring between the gut and the brain (O'Mahoney et al., 2014; Mogilevski, 2021). The resulting microbiome alterations and accompanying GI inflammation lead to excessive T-junction permeability and global inflammation of the interstitial process. The pro-inflammatory kynurenic pathway has been reported as a key aspect of the depression process and reroutes 5-HTP into quinolinic acid and reduces serotonin levels (Ozturk et al., 2021). Recent research suggests the effect of SSRIs in depression is mainly due to their anti-inflammatory properties more than their impact on serotonin reuptake in the synapse. As individuals shift in and out of depression intensity and related alpha asymmetry, it is common for them to become re-sensitized to their fears and frustrations moving them into more suicidal, irritable, and angry states. It is well known that when individuals with depression come down from a dose of Ritalin and related isomers, they commonly have irritability, anger, and rage episodes. A large percentage of children who were clients at our office had parents continually complaining of oppositional and angry outbursts on the ride home from school as the last dose of school related stimulant wore off.

The ISI Profile of Trauma and How to Implement

The Interactive Self Inventory (ISI) is a 16-dimension socio-emotional assessment designed to be used in conjunction with brainmapping. It divides behavior into eight dimensions of Approach behaviors and eight dimensions of Avoidance behaviors linked to measures of anxiety and depression. It is extremely useful for identifying features of trauma and the social emotional patterns through which they are expressed and relating them to the EEG.

The psychophysiological aspects of anxiety and depression are defined in terms of arousal and inhibition, anxiety being represented as overarousal and depression as excessive inhibition. This is done to correlate them with the social dimensions of approach and avoidance which are in turn defined in terms of beta and alpha asymmetry. These psychophysiological dimensions of arousal and inhibition are labeled as anxiety and depression in the ISI in order to assist in cross-disciplinary interpretation of the ISI. Anxiety and depression are the psychological terms that can be translated into the psychophysiological terms of arousal and inhibition. The difference in terms is related to the difference in paradigms in the same manner as the psychological term "complex trauma" is another way of characterizing "toxic stress" depending on your disciplinary perspective. From a psychophysiological perspective they represent the gas pedal and brakes of the ascending reticular system as it relates to the sympathetic and parasympathetic systems of the CNS. They also reflect the vagal system activity as well as it is defined and presented in opponent process vagal theory (Craig, 2005). The approach and avoidance measures in the ISI are, as I mentioned, linked to the EEG literature through the metrics of asymmetry (Soutar, 2017a; Solomon et al., 2014). In this manner we can couple the concepts of anxiety and depression with arousal and inhibition and with EEG asymmetry as well as with approach and avoidance. By doing this we create an assessment that is truly bio-psychosocial. The approach and avoidance dimensions then become the meta dimensions of the ISI with the other dimensions representing subdimensions. For more on the statistical procedures and dimensional concepts used to construct the ISI see my paper on the Interactive Self Inventory (Soutar, 2008).

A significant portion of the population with complex trauma experience deep concerns regarding how they appear to others because of the intensity of their own shame. They feel exposed and readable to others and may adopt a very flat affect to hide their emotions and protect themselves. This shows up in especially high scores in the Inhibited dimension on the ISI (not to be confused with the psychophysiological dimension of inhibition related to depression). In addition, they have elevated scores in the dimension of Perfectionism and Competitive which we found manifests as shame and internal self-derogation.

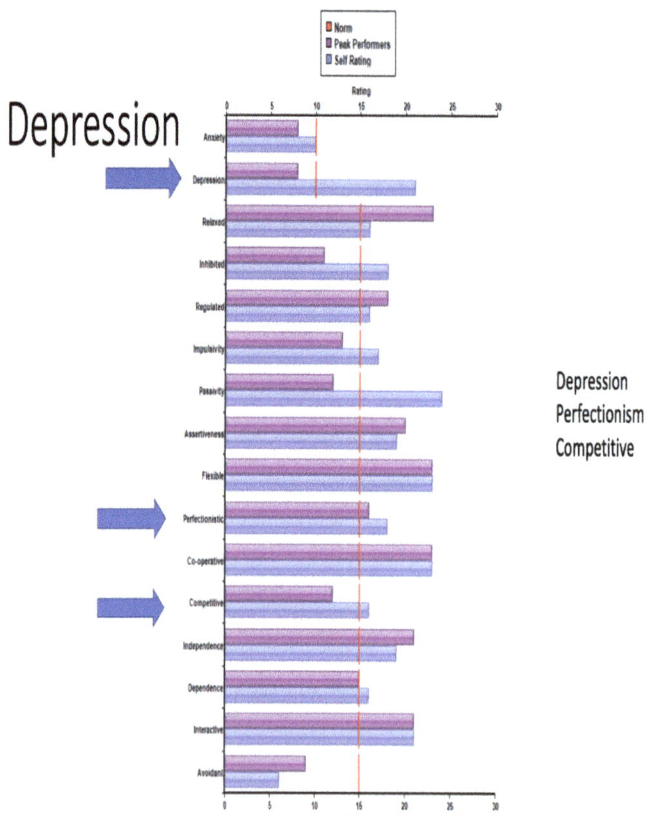

The main forms this self-derogation in terms of cognitive activities were as listed below and they reflect classic dimensions of automatic thoughts identified in CBT theory as defined by Judith Beck (1995).

All-or nothing thinking

Catastrophizing

Disqualifying

Emotional reasoning

Labeling

Magnification

Mental filter

Mind reading

Overgeneralization

Personalization

Shouldism

Tunnel Vision

When investigating anxiety and its associated psychophysiological state of chronic hyperarousal, I found strong statistical correlations with the socioemotional dimensions of Perfectionism and Impulsivity in the ISI. The impulsivity as defined here is usually related to overwhelming emotional episodes that overcome inhibitory prefrontal cortical mechanisms resulting in impulsive behaviors. This underactivation and hypoherence resulting in this inhibitory deficit has been observed in the frontal networks in various imaging studies including quantitative EEG investigations (Nicholson et al., 2020). That impulsivity is also reflected in the output graph above. In this scenario fear in the form of excessive subcortical inputs from the amygdala reduces frontal inhibitory control of behavior.

In addition to the ISI, we provide a checklist of questions that has been correlated with the Test of Variables of Attention (TOVA), The Beck Inventories, the Microcog neurocognitive battery, and the Stroop Interference Test. This is offered to get a quick and dirty assessment of anxiety, depression, memory, attention, and sensory integration, and it has proven to be actually quite accurate through statistical analysis. We call this the Cognitive Emotional Checklist (CEC) and we have one version for adults and one version for children. For those interested in the technical aspects there is a paper on it in the Help section of the NewMind Maps or Reports website along with the paper on the ISI. The answers to the items on this checklist are correlated with the EEG on dashboard section of the map output to confirm or disconfirms the items subscribed to by the client.

A large number of children in my clinic who typically scored poorly on the Test of Variables of Attention (TOVA), also had indications on the TOVA of anxiety and or depression. These children also scored significantly on the Cognitive Emotional Checklist (CEC) in the dimensions of anxiety, depression as well as the complaints of poor attention in the CEC attention rating scale when parents rated them. Anxiety and depression have a profound impact on cognitive function. Often when the emotional components are somewhat resolved and integrated, cognitive function automatically improves. This is why we recommend dealing with emotional issues first before taking cognitive assessments too seriously, then test for cognitive function a second time after anxiety and depression diminishes.

Cognitive / Emotional Checklist Assessment

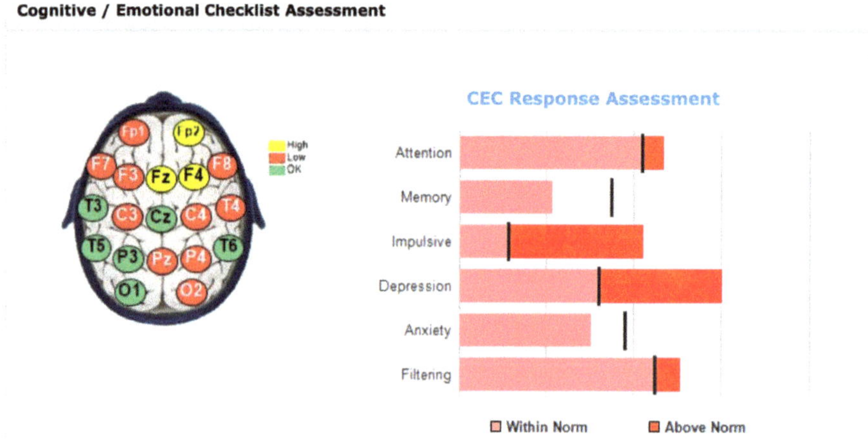

So clinicians should look for and expect some unique variations of common dysfunctional mental activity embracing several of these dimensions in the ISI. As the scores in these dimensions of the ISI increase, so do the frequency and severity of these negative automatic thoughts. It is of significant value to help clients watch for these negative thought patterns as they do HRV training prior to neurofeedback sessions. The manifestations of these negative self-talk patterns come in waves and correlate with episodes of poor heart rate coherence during HRV training. This helps sensitize clients to their areas of most significant emotional challenge and how they can identify and integrate negative episodes of dark thoughts. By allowing these thoughts to enter their conscious awareness rather than operate automatically at an involuntary level, they can take active measures to manage them as well as calm themselves physiologically. In this process clients come to tacitly understand how they habituated to these thoughts and feelings as these thoughts and feelings drop below the threshold of awareness and their experience. This activity then becomes a confirmatory process of empowerment as it can be verified and altered in terms of its impact physiologically as measured with biofeedback and neurofeedback equipment. Clients experience the power of automatic thoughts and emotions to directly impact their body from moment to moment.

Those scoring high in Passivity are primarily conflict avoidant, which is another typical aspect of trauma as well as Codependency. In fact, trauma and addictive behavior, as previously established, are commonly found together. Alcoholism levels in traumatized veterans returning home is a common theme in VA efforts to manage veteran suicides etc. Pleasing others to remain safe can make some clients impressively sensitive to other's emotional states. The empathic powers of many therapists come from a history of trauma that has been overcome through counseling and their own integration. Hypervigilance in conjunction with highly developed mirror neurons in this instance can be turned into a skill that assists others in their healing. Clients with high Passivity scores will have learned in their family of origin to suppress their own emotions in order to better monitor the emotions of the dominant disordered individual driving the family system. In the case of alcoholism this is in the direction of a rigidly disengaged state commonly found in families with addiction related issues (Soutar, 1993). When Dependent dimensions measures are high in conjunction with the elevated ISI scores in the Passive dimension, then an addictive process is likely present in the socializing family of origin.

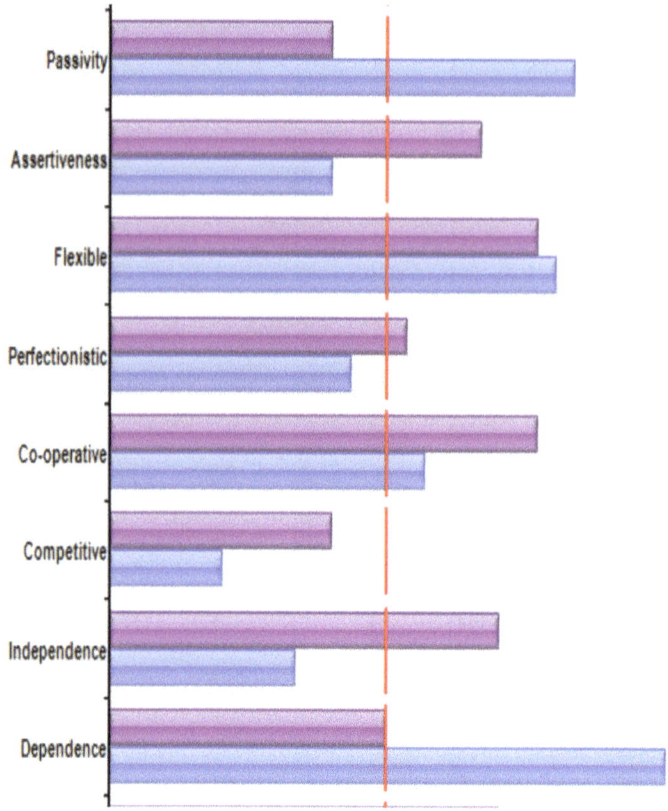

Those scoring high on the ISI dimension of Avoidance are usually emotionally numbing and isolating socially to avoid triggering stimuli. From the literature on complex trauma previously discussed, these triggers can be multitudinous and difficult for clients to articulate, and consequently social isolation is a simple general solution. This can evolve into agoraphobia or just diminished emotional engagement and interaction. Focusing excessively on work tasks is often an optimal professional method of self-distraction. I have encountered many successful professionals with this pattern. When low scores in approach behaviors are also involved, it usually indicates that individuals are overly self-absorbed with distracting internal negative self-processing and tend to personalize any challenges to their opinions or social interaction patterns. They will typically have high levels of posterior DMN activation which can manifest as elevated alpha or beta depending on the negativity of their self-talk and bracing.

All of the profiles above manifest varying degrees of psychological anxiety and depression along with the various physiological symptoms. These two dimensions appear in most deviances in the neurometric dimensions seen in the qEEG. These patterns are reflected in their neurological network patterns as well as the growth and development of various Brodmann areas of the brain as they relate to these networks. Different profiles will have different qEEG patterns that will be reflected in unique deviances from the normative distribution. These patterns change with new insights and new behaviors. The changes can be observed in pre-post maps regardless of whether the client is receiving fast or slow wave training. A most interesting point to be made here is that constant engagement with the AT crossover state alone can alter both the ISI and the qEEG significantly and move them closer to the statistical norm. This in itself is a strong argument that NFB affects the brain as a system and can be implemented locally with fast wave training or globally with slow wave training such as AT and alpha frequency and synchrony training.

The Social Denial of Trauma and Its Impact on The Individual

The authors of the DSM-III in 1980 openly acknowledged that it was so imprecise "it should never be used for forensic or insurance purposes" (van der Kolk, 2014, p. 29). This warning was ignored and the rest is history we have explored. This willful ignorance, however, is characteristic of how whole societies avoid confronting unpleasant realities, whether it is global warming or the side effects of food additives or pollution. One might wonder if consumerism and the associated poor skills with delayed gratification are an epidemic. Mental health issues are clearly a reflection of the social order of the day. The history of psychology alone is very instructive with respect to this issue. The position paper for the American Academy of Pediatrics was quite clear on this inevitable conclusion that we were ignoring many other variables contributing to health issues reflecting a general social denial process.

"Although new discoveries in basic science, clinical subspecialties, and high-technology medical interventions continue to advance our capacity to treat patients who are ill, there is growing appreciation that a successful well-care system must expand its scope beyond the traditional realm of individualized, clinical practice to address the complex social, economic, cultural, environmental, and developmental influences that lead to population-based health disparities and unsustainable medical care expenditures" (Shonkoff & Garner, 2012).

The denial even takes on a structural pattern socially that reflects entrenched values and beliefs that are both cultural and subcultural. Unhealthy communities have too many fast food franchises and are food deserts. They have excessive numbers of liquor stores as well as diminished opportunity for physical activity due to higher risk of exposure to high crime rates that also disrupt social networks that stabilize communities.

An ecological model clarifies the multiple etiologies and causal vectors that converge to undermine psychological development as well as amplify health risks. Longitudinal studies show that changes in a child's ecology have measurable effects on the child's developmental trajectory. They influence educational achievement, economic productivity, health status, and longevity. As Shonkoff and Garner note, "There is strong scientific consensus that ecological context modulates the expression of one's genotype." This involves the expression of some behaviors and inhibition of others. The cumulative stress burden over time generates an "allostatic load" that impacts the entire physiology and growth structural patterns of the human organism.

The story that has emerged is so well-documented and has been for so long that to even bothering to cite it is an insult to its enormity. The politics of economic production heavily influence and distort the distribution of accurate information to consumers. Uninformed consumers often rely on contaminated or substandard resources that define family social context and undermine family health. This in turn defines the spectrum of behavioral repertoires available to family systems as they attempt to respond to top-down social pressures. The social order, as sociologists are fond to say, is truly external and coercive. Commerce and profit define resources to be used in the production of food and food additives that alter health and behavior. The basic resource of water is defined by industrial needs as most water is not used for residential purposes. The distribution of wealth and cost of keeping water potable has profound consequences on family health as recent new stories have made publicly and agonizingly clear. Industrial pollutants in air and water affect the entire physiology and especially the brain and CNS. Heavy metals such as mercury and lead contaminants have profound impact on gestation and early childhood neurological development. Dioxin, plastics, and glyphosate, just to get started, are permeating our adipose tissue and destine us for a lifetime of health issues. All of this impacts the brain and behavior. The radical rise in neurological issues such as autism is alone sufficient testimony to our dilemma. All of these factors contribute profoundly to human trauma and its acceleration and transmission. The manifestation of trauma could be used as a metric for social health as the ACE study demonstrates.

Based on the foregoing it would seem incumbent upon therapists to take these issues into account when taking client histories as well as for researchers to extend their consideration of confounding variables, but few do. Trauma is a social phenomenon as well as a psychological and physiological phenomenon. Hence complex trauma and toxic stress are something that is to be more effectively managed than eliminated. This is why clinical efforts to resolve it can take 10 or 20 years. A more functional approach to medicine and complex trauma is required and at NewMind we have adopted a Functional Neurofeedback paradigm to cope with it.

Chapter 5

The Physiology of Trauma

The Role of The Physical Body

The physical body is the ledger accounting for the impact of trauma on the human nervous system and provides a symptomatic history of each event that transpired. The body is the score card as van der Kolk (2014) characterizes it. The physical body bears the scars of battles past but these wounds do not lay quietly sleeping. The body is a dynamic allostatic organic system that must constantly perform a balancing act to maintain baseline functioning. There are constant checks and balances to be made and the juggling act in itself becomes more stressful as time passes and each new threat challenges it. The dynamics create pressure on different networks, endocrine and organ systems that are very real in their consequences at the organic level. In the extensive animal research on stress in the field of psychology in the late twentieth century it was found that most lab animals, mammals, died of gastrointestinal lesions as in the executive monkey experiments.

The fact that these mammals, including the famous executive monkeys, died of stress to their central nervous system and not physical trauma reveals several basic truths for humans. We too can suffer poor health and die from stress. As a society, the individual is buffered to some degree from the physical ecological stress but that ecological stress becomes translated into group dynamics that generate equivalent or worse stress. We are protected from the wind, the cold, and the wild predators, as well as from starvation, but it comes at a cost of group interaction dynamics. As a species we define how we experience that stress. We may deploy pesticides and herbicides in agriculture to protect us from starvation with high yield harvests but that may translate into deposits of carcinogenic substances in our adipose tissue. This in turn generates human drama and stress in the form of healthcare costs, lives lost, and a host of other sociological factors. The Book Why Zebras Don't Get Ulcers by one of the leading investigators on stress at UCLA Berkley's Robert Sapolsky (1999; 2004) thoroughly outlines the entire phenomenon based on very high-quality research.

War and crime are other social phenomena that arise from this translation and that in turn generates the PTSD and complex trauma that begets the dysfunctional transgenerational behavior that continues its impact through successive cohorts. This translates into social dynamics that generate dominant hierarchies that create considerable suffering. Sapolsky found in his research that once the dominant malignant benefactors, the dominant males, of the baboon troops he was studying were wiped out by a plague, the social structure of the successive cohorts displayed more sociality and less stress as measured by decades of carefully logged blood samples. When these insights were applied to the White Hall studies on hierarchy and stress in England it became clear that humans have their own version of this ancient pattern of behavior. The ACE study merely confirmed it again from another perspective. Stress kills, and it does so quietly, in a manner of speaking. Stress feeds and nurtures emotional trauma which in turn eventually translates into complex trauma.

Stress response intensity varies with many factors including age, gender, personality, physical and mental health as well as past experience (Sharma, 2018). When fear escalates, the adrenal glands, responding to inputs from the amygdala, generate adrenaline and also trigger a global endocrine response in the hypothalamic pituitary axis or (HPA). This fear response also generates high levels of norepinephrine and major changes in the brain. Cortisol is generated in the body to calm the system back down after a stressful event but too much constant stress exhausts the body's resources including the availability of cortisol. This is one of the many ways that constant fear and stress changes the body permanently. Cortisol is chronically low on average in those who have PTSD and complex trauma (Sherin & Nemeroff, 2011). Since cortisol suppresses the stress response chemicals, the body does not return to baseline in PTSD. It is always on high alert for danger and in biofeedback we call that chronic readiness bracing.

Bracing, a common response to chronic stress and trauma, can often be observed in the practice of biofeedback as globally elevated muscle tension. This can be empirically measured using body sensors that amplify the associated electrical activity of bracing and track it using computer screen graphics. The majority of people who have complex trauma commonly display high levels of what we call EMG, or electromyographic activity or simply muscle tension. The scalp, in particular, is an excellent place to measure this activity. When we measure brainwave activity, we also receive this EMG information. In fact, it is a much stronger signal than the EEG and often causes interference, especially above 21 Hz (Soutar, 2023). When training clients with neurofeedback we can monitor the real time changes in bracing and observe how most individuals with complex trauma experience increased scalp tension when they relax too much out of the ordinary due to the neurofeedback. We can also see EMG increases when they begin to achieve crossovers and enter a state where they may reexperience the trauma through emotional memories and visual memories etc. prior to integrating them. This pattern of episodic bracing becomes a powerful clinical tool for monitoring integration, which I will explore more in the chapter on implementing AT training.

In my book, The Automatic Self, I describe my first experience with this technology while in graduate school. I was being monitored in a clinic with biofeedback equipment and was asked if I felt relaxed. When I expressed that I was quite relaxed they showed me the display screen on the computer that empirically showed I was not even close. I had habituated to my own bracing, which most people do, and was completely unaware of how tense I actually was at the time. It is quite a shock to tacitly experience this phenomenon. You begin to understand how well and how quietly the body keeps the score. Most of us are completely unaware of the cumulative effects of stress on our body because we adapt to our own bracing over time until we are totally unaware of it. The majority of the clients I have done brainmaps on over the last two decades have significant EMG which interferes with getting a clean EEG record. It is so common that it has even likely distorted the qEEG database norms more than we realize (Soutar, 2023).

I remember one young man, who was an amazing example of this, and a client of one of my associates when I had an office in Scottsdale, Arizona. My associate asked me to bring along my biofeedback equipment to measure the bracing of this highly traumatized individual. She was intrigued by my technology and wanted to observe it in action in the clinical setting. When we had him hooked up, we found he had a very high baseline and then she proceeded to discuss some of his memories with him. As he talked, you could see his EMG go right off the chart. He indicated afterward that he did not experience any change in his tension level, but when we asked him to try and consciously tense up enough to duplicate what we had recorded, he could not do it. I knew this would be the case because I had tried myself and with many of my clients but I had never seen such an extreme case before. The human capacity to unconsciously adapt to extreme states of physiology is truly amazing.

Maier and Seligman in 1967 conducted some famous experiments with dogs that likely would not pass muster with a contemporary ethics approval process (Maier & Seligman, 2016). They placed dogs in a cage where they would receive a mild electric shock unless they pushed a lever. As the dogs learned to do this the experimenters yoked them to a second group of dogs in a manner that made it so that for the second group there was no escape from the shock using the lever. In a sense they created traumatized dogs in this group where the lever did not work. When they placed the dogs in a cage where they could leap over a barrier to escape the shock, only the dogs who had previously had a working lever attempted to jump the barrier. The traumatized dogs did not attempt to jump the barrier. They had lost hope of escaping. These dogs had to be dragged over the barrier in the cage to relearn that escape was possible. The theory was that they had learned to be helpless. Although more recent experiments by Maier (Maier & Seligman, 2016) have more finely qualified these findings, for our purposes the significance of the outcome remains the same.

It is during the crossover state in AT training that clients learn they can experience the exposure of remembering without the shock of extreme fear and panic overwhelming them. It is at this point we also notice that clients begin to spontaneously emit novel behaviors in circumstances that previously would have elicited automatic thoughts, emotions, and behaviors based on an extreme fear response. They discover they can escape and even manipulate the situation to their advantage and thus experience a moment of empowerment. They move from a rigid reactive state involving habitual response to an event to a spontaneously creative response reflecting a newfound level of neuroplasticity. This is one aspect of the enhanced creativity that the Greens sought to achieve through AT training. Another aspect is the ability to solve problems that at first resisted solution, as in Elmer's ability to solve mathematical problems in physics in the AT crossover state.

Hans Selye developed the stress theory of disease in 1946 which became known as the general adaptation syndrome (GAS) that found its most sophisticated expression and confirmation in the ACE study (Shonkoff & Garner, 2012). Fundamental to the theory is that unrelenting stress progressively overdrives and exhausts the biological system making it vulnerable to disease and death. In his work he identified three stages of the stress response known as alarm, resistance, and exhaustion. The GAS theory is one of the key concepts in psychophysiology. It identifies stereotypical and expected patterns of physiological response to stress that are well-documented in the research and which will be summarized below. In addition, correlations can be found, as described above, in the psychological response to stress in Maier and Seligman's experiments with dogs demonstrating that prolonged stress with no escape leads to a state in dogs that mimics human depression.

Based on the preceding, chronic hyperarousal in the form of anxiety leading to depression and chronic inflammation and then ending in death for mammals, and humans in particular, is a pretty well established pattern. This pattern also appears to be robustly reflected in the oxidative stress process emerging from inflammation in the brain and, as it turns out, quite observable in patterns of the qEEG, if you know what you are looking for in that assessment, as presented in the next chapter (Soutar et al., 2017a,b). As we have seen, extreme forms of stress typically emerge in developmental (toxic stress) and complex trauma.

In reviewing the literature on how the body responds to stress, there is also an observable sequence of stages as in the qEEG. In this response sequence of the body to physical or social distress a clear general, but very complex in detail, progression of events occur. The complexity of the details has lead to very narrowly confined disciplinary investigations that have obscured the general picture. Interdisciplinary analysis brings to light a basic pattern once the noise is filtered out. In our 2013 presentation at ISNR, I proposed the idea that fast and slow oxidizers may actually be two aspects of a continuum in the same manner as anxiety and depression as described previously. In addition, I explored with a statistical model that anxiety and depression are behavioral manifestations that exist in a bi-directional relationship with the physiology of this pattern. This was finally summarized in a multi-clinic project described in my paper on oxidative stress (Soutar et al., 2017a). These two phases are defined as follows:

Fast oxidation is defined on a properly performed hair mineral analysis when the calcium/potassium ratio is less than about 4 and when the sodium/magnesium ratio is greater than about 4.17. The lower the calcium/potassium ratio or the higher the sodium/magnesium ratio, the faster the oxidation rate.

Slow oxidation is defined as a calcium/potassium ratio greater than about 4 and a sodium/magnesium ratio less than about 4.17. The higher the calcium/potassium ratio or the lower the sodium/magnesium ratio, the slower the oxidation rate.

Anxiety is characterized as fast oxidation and part of the alarm phase of distress involving elevated levels of adrenaline and overactive adrenals, rising cortisol and elevated glucocorticoids. It is well-established that there is also a comprehensive mobilization of the endocrine system through the hypothalamic pituitary axis (HPA) as well as significant changes in the microbiome. A review of my clinical data showed a .719 correlation between adrenal symptoms on our physiological assessment questionnaire and reports of features of anxiety on our Interactive Self Inventory assessment.

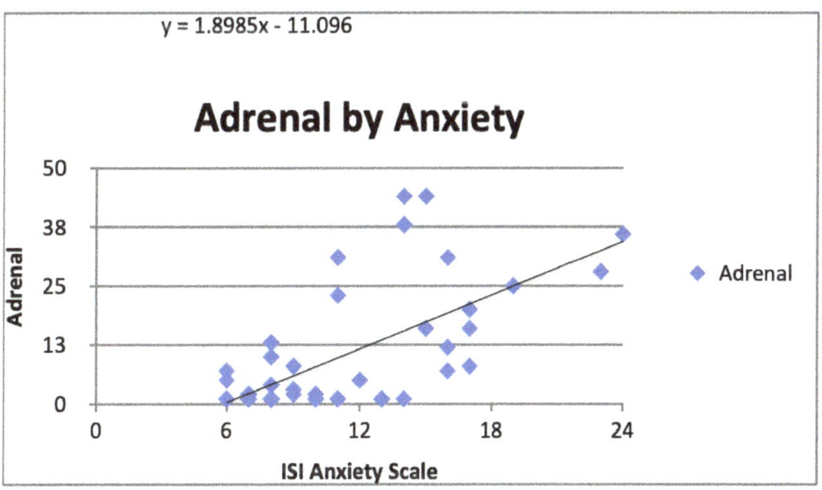

Depression is an advanced stage of anxiety presenting as an exhaustion process often reflected in thyroid depletion and dysregulation. In addition, there is increasing inflammatory response in the brain to the build of glutamate in the brain (Wang & Michaelis, 2010). Similar data from the clinic shows a .574 correlation between symptoms reported on the physiological checklist and the ISI.

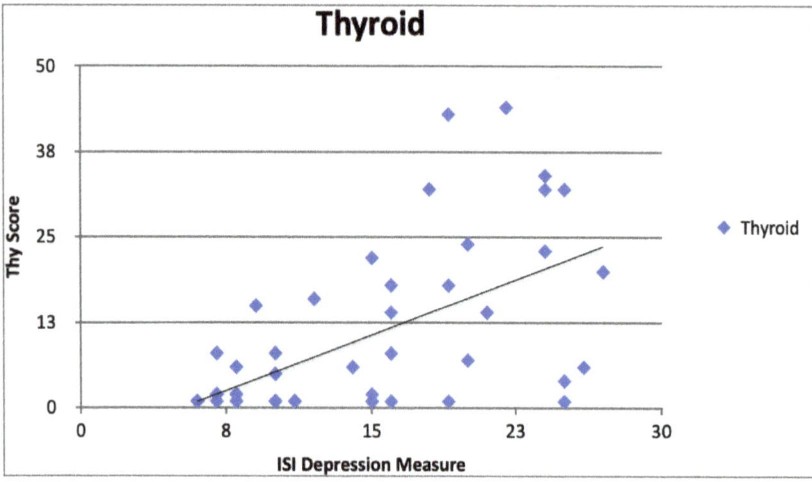

Variation in the response of the body to chronic distress results in less contribution specifically to reductions in thyroid function and a lower correlation than anxiety and adrenals but the correlation is still very robust by statistical standard. Exhaustion is reflected in the body in a wider array of symptoms than the initial anxiety and alarm stage. Thyroid T3 plays a key role in reuptake of norepinephrine and dopamine and T3 would naturally parallel changes in these major neurotransmitter levels (Gilbert & Zoeller, 2010). A variety of toxins can also contribute to the manifestation of thyroid deficits that can contaminate analysis.

Some individuals may initially respond with symptoms more related to hyperthyroid which can also vacillate between high and low thyroid function as the organism reaches exhaustion and slips into permanent low thyroid function.

The Initial Stages of Stress

The autonomic nervous system and the HPA are the two major stress signaling pathways which interact with the CNS to generate an adaptive response to stress (Seiler et al., 2020; Tsigos et al., 2020). The stress response involves a complex interconnected neuroendocrine, cellular, and molecular infrastructure that defines the stress system located in the central and peripheral CNS. Growth metabolism, reproduction, and immune function can all be negatively impacted including behavior and personality development. The stress response evolves uniquely at the individual level through the prenatal period, infancy, childhood, and adolescence and into adulthood. Consequently, each individual's stress response history shapes their physiology at both the structural and process level based upon their history. Their ability to respond effectively and recover is based upon their own and their family's social resources. Each person is different in their response to the triggers of stress and the intensity and character of their response varies greatly. Energy is redirected away from many routine physiological activities, such as digestion, during stress periods toward other activities critical to immediate survival. A prolonged adaptive stress response, defined by slow psychophysiological recovery, can become maladaptive and destructive to the system, contributing significantly to inflammation and disease. Stress also has a positive side, which is referred to as eustress, and can be rewarding in some instances and can help reset the physiological system. An example might be a wedding or a roller coaster ride.

Appetite, sleep, energy levels and cognitive function can be significantly affected by stress. Glucocorticoids, generated by the stress response, can affect secretion rates of specific proteins and alter the electrical potential of neuronal cells as well as mitochondrial function and energy metabolism of the brain (Tsigos et al., 2020). We often see this in the qEEG in the first stages as elevated high frequency activity and then eventually as reduced high frequency activity as resources become depleted. This is an indicator that the brain is running low on energy as it takes significantly more energy to produce high frequency waves than low frequency waves. Glucocorticoids also play a crucial role in basal HPA activity managing the boundaries of the stress response and termination. They influence ultradian cortisol pulsatility which in turn impacts gene expression and the phenotype of target cells. This suggests deep and enduring configurations in physiology and development are likely outcomes.

Stress actually defines our physiology during development. We grow an anxious body and an anxious brain—they don't suddenly develop that state. This helps explain the intransigent nature of complex trauma. It also suggests that resolution would involve adaptive growth to redefine the system that would require years of change with respect to treatment. In essence, the stress response is global in its impact on the body and organ systems. The stress response sets the arousal level, which is moderated by the reticular activation system (RAS) in the brain stem (recall that traumatized individuals have higher arousal levels). The hippocampus plays a key role in regulating and shutting down the HPA response because of its inhibitory relationship with the amygdala. Consequently, hippocampal degradation due to prolonged stress can reduce its ability to inhibit amygdalic activity. The core body temperature is typically elevated due to stress to the paraventricular nucleus (PVN)-corticotropin-releasing hormone (CRH) neurons systems and locus coeruleus (LC) norepinephrine system (NE; LC/NE) noradrenergic activity. These systems characterize and link stressors to previous experience (Wang & Michaelis, 2010).

Glucocorticoids directly affect appetite regulation and stress induced eating, and yet ironically at the same time some configurations of the CRH physiological stress response can result in anorexia. The central clock mechanism of the body revolves around the circadian glucocorticoid rhythm that synchronizes peripheral organ clocks that maintain their own autonomy. So, this rhythm is central to the entire body's coordinated response. Chronic elevated glucocorticoid levels dysregulate this clock and organ synchronization.

Reproductive hormones are also suppressed during periods of high stress such as gonadotropin-releasing hormone and steroidogenesis (cholesterol is converted to steroid hormones). Chronic HPA activation and related elevations of glucocorticoids are associated with decreased production of thyroid-stimulating hormone (TSH) which inhibits conversion of the thyroid homone T4 to its critically important form of T3 for neuronal function. Cortisol excess can have the same effect. Glucocorticoids typically enhance the intake of carbohydrates and fats, but acute stress suppresses appetite. Childhood stress contributes significantly to obesity. Sexually abused girls also exhibit chronic HPA axis activation. There are links between chronic painful gastrointestinal disorders and a history of physical and psychological abuse.

The amygdala nuclei in the brain are the main centers for CNS fear-related behaviors and central to the activation and retrieval and emotional analysis of any stress related stimulus. Thus, the amygdala plays a key role in retrieval of trauma memories and their consolidation as well as the response of physiology to that retrieval. As we shall see, this is a key reason for the inability of clients to integrate traumatic memories. Wang et al. (2019) notes,

"As such, the central components of the stress system are located in the hypothalamus and the brainstem and include the parvocellular corticotropin-releasing hormone (CRH) and arginine-vasopressin (AVP) neurons of the paraventricular nuclei (PVN) of the hypothalamus, and the CRH neurons of the paragigantocellular and parabranchial nuclei of the medulla, as well as the locus coeruleus (LC) and other catecholaminergic, norepinephrine (NE)-synthesizing cell groups of the medulla and pons (central sympathetic nervous system) [1–4]. The peripheral limbs of the hypothalamic-pituitary-adrenal (HPA) axis, together with the efferent sympathetic/adrenomedullary system, constitute the peripheral components of this interconnected system."

Once excited by the amygdala, the hypothalamus is the first brain center to initiate action in the HPA. There is a complex response here involving the sympathetico-adreno-medullar axis (SAM) and the HPA axis in conjunction with an immune system response (Chu et al., 2022; Sharma, 2018). There are fundamentally two general responses. There is a slow response from the HPA and a fast response from the SAM. The fast response results in increased levels of norepinephrine (NE) and epinephrine from the adrenal medulla and increased NE secretions from the sympathetic nerves. This results in increased activity in the right hemisphere of the brain, which is more highly responsive to norepinephrine than the left side. You can often see this fast response during training sessions when clients are watching a tense section of a video that triggers them.

During stress there is also an interaction with receptors on smooth muscles and other organs of the body. The smooth muscle contractions of the arterial system show up as changes in the peripheral vascularity measures of biofeedback such as hand temperature. So, there is an immediate general impact on major organ systems. This impact of the contraction of the smooth muscle cells involving vasoconstriction also results in increased blood pressure, is associated with increased heart rate, and affects skeletal muscle blood flow. There is increased sodium retention, increased glucose levels, lipolysis (the breakdown of fats into fatty acids and glycerol), increased urea, increased oxygen consumption and thermogenesis. There is also reduced intestinal motility (constipation), cutaneous vasoconstriction (in case you bleed), bronchiolar dilation (to get more oxygen), and suppression of insulin (and appetite) with a rise in blood sugar. The bronchiolar dilation is in concert with more rapid and shallow breathing that shows up as reduced HRV coherence. Heart rate spikes at the same time. This is why HRV can be such a useful and sensitive measure of internal trauma-based triggers during AT training. The SAM system involving the adrenals has immediate behavioral consequences with enhanced arousal, alertness, vigilance, focused attention and analgesia (in case of physical trauma). This is all standard sympathetic activity for most individuals with fairly common long-term effects modulated by DNA variation which is part of the diathesis stress model. Our genetic heritage has a strong impact on how stress impacts us, and different aspects of our physiology are most vulnerable according to that heritage.

The slow response from the HPA involves another process wherein the release of corticotropin-releasing hormone (CRH) impacts peripheral tissue and the subcortical structures of the brain. CRH also stimulates the pituitary gland to release ACTH in the bloodstream which in turn stimulates the adrenals, the famous adrenaline rush, to secrete glucocorticoid hormones such as cortisol and these glucocorticoid hormones build up over time to regulate the process.

Stress affects all organ systems of the body, so the body does register and keep the score. Cardiovascular, respiratory, endocrine, gastrointestinal, nervous, muscular, and reproductive systems are all impacted and can be subject to reactivation during periods of flashback memories. It is even likely that the sequence of alarm arousal that initiated the original response is reproduced faithfully. This aspect of complex trauma is often overlooked but is well-studied under the rubric of structural dissociation as developed by Pat Ogden (Steele at al, 2005). It is important to keep in mind that stress not only affects GI motility but additionally nutrient absorption. This can, in turn, because of the bi-directional signaling of the gut-brain axis, have a long-term effect on ATP production at the neuronal level. Without sufficient minerals and other nutrients, the sodium potassium pumps cannot work efficiently either.

The immediate consequence of stress is to increase the adrenaline and norepinephrine levels in the body. Glucocorticoid levels rise and, through hippocampal mechanisms, reduce immune response, which has been common knowledge since the 1990s. As previously mentioned, cortisol levels rise in response, in order to temper CNS arousal. Unfortunately, rising cortisol levels can damage hippocampal cells and diminish short term memory functions. The GI system slows down, and the GI environment becomes more acidic, shifting the microbiome bacterial profile and increasing GI inflammation. Adenosine accumulates from lack of sleep and creates slowing in the EEG, involving reduced cognitive function and brain fog (Walker, 2017). This can be seen in the qEEG as reduced alpha frequency activity in the dominant frequency metrics and often increased alpha amplitude as well. This increases cognitive errors, resulting in mistakes that further aggravate adrenal output. Another negative feedback loop takes hold, and social accuracy diminishes, as irritability and emotional instability increase. Adenosine also undermines the nocturnal peak of melatonin that provides the body clock with signals that it is time to go to sleep. Insomnia and general sleep dysregulation and associated parasomnias emerge.

In reviewing the literature, a general pattern emerges, of key mineral deficits, hormonal imbalances, and other resource deficits, that is fairly stereotypical of the human physiological response to stress. Each individual has unique differences in the details, but these deficits and imbalances alone can be critical for general physiological functioning. The figure below summarizes the process.

Initially, in the alarm phase, calcium and magnesium become depleted while sodium, cortisol and glucose levels rise. Aldosterone increases, promoting sodium retention and potassium reduction. Growth hormone (GH), a peptide from the pituitary gland increases the concentrations of glucose and fatty acids, while the thyroid releases thyroxin and triiodothyronine, increasing basal metabolism and heart rate. These can all become chronic features of enduring metabolic states during chronic stress. There is an increase in cortical activation resulting in initial elevated levels of beta frequencies and diminishing levels of alpha activity.

During the resistance phase, potassium decreases along with blood sugar, insulin and cortisol. At this point there is a major GI PH and floral shift along with DHA, B12, B6, and folic acid deficits in conjunction with increasing inflammation. Energy levels in the brain are compromised with progressive loss of ATP. Beta activity diminishes in amplitude, while alpha activity begins to increase, as the brain moves progressively into stand-by mode.

Moving into the exhaustion stage, selenium drops as cortisol and thyroid T3 become depleted. This compromises neuronal ion pumps even more and alpha frequencies become even higher in amplitude, while the frequency distribution of alpha begins to emphasize slow alpha frequencies.

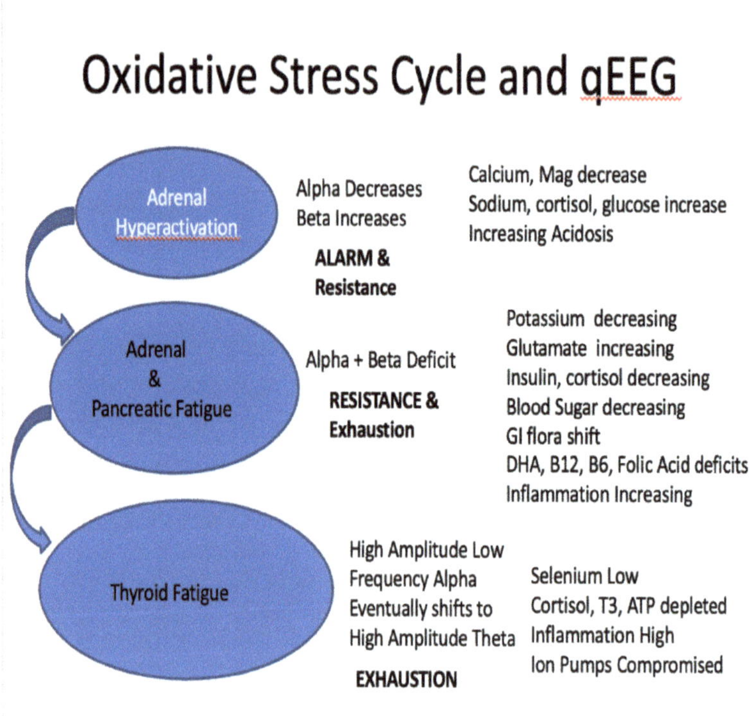

As the microbiome shifts progressively into a greater inflammatory phase, it triggers a bidirectional communication between white blood cells in the intestinal lining and white blood cells in the brain. A pro-inflammatory response in the brain initiates a conversion of 5-HTP into quinolinic acid along the kynurenic pathways and resulting in reductions in brain serotonin (O'Mahoney et al., 2015). At the same time, GI inflammation increases tight-junction spacing in the intestinal lining, releasing protein leakage into the interstitial spaces and raising zonulin levels. Often an over-reaction of cytokine activity dramatically increases interstitial inflammation, which affects many other organ systems and their hormonal signaling mechanisms and effective co-ordination with the HPA. Stress eating often emerges, due to dysregulation of the GI signaling through vagal pathways to the hunger centers of the brain. Chronic tight-junction issues trigger excess releases of lipopolysaccharides that significantly burden liver detoxification processes (An et al., 2022). The consequences of this general pattern of brain and GI interaction is that it progressively impacts the entire body. This physiological process is at the same time an intimate part of the psychological process that distorts behavior. The research that unveils the deep interconnectedness of these psychological and physiological process are a testimony to the imperative importance of addressing contemporaneously each level of its manifestation.

Pathophysiology

The features of chronic stress, as opposed to acute stress, are likely to be more developed in complex trauma. As argued previously, chronic stress generates enduring physiological and behavioral changes that may be difficult to reverse as these changes become permanent features of structure and function. Blood pressure and heart rate may become permanently elevated as does chronic muscle bracing, which can often be localized based on the trauma experience. The gastrointestinal tract may be permanently affected through decreased local blood flow, as well as inflammation. The immune system may become weakened, and sleep disorders emerge. Cortisol dysregulation is correlated with poor sleep quality and daytime drowsiness (Hirotsu et al., 2015). Addison's disease and Cushing syndrome may emerge. Generally, cortisol levels drop after acute stress as the body attempts to regain balance and stability. If stress continues for extended periods of time, the body continues to secrete stress hormones on a regular basis.

Chronic stress may increase cholesterol and plays a role in the accumulation of abdominal fat, which is a prominent storage point for toxins absorbed from food as well as environmental sources. So, there is a synergy between physical toxins and psychological toxins. One amplifies the other. Chronic inflammation is typically an outcome of chronic stress, leading to or playing an important role in a variety of chronic illnesses, including cancer, type 2 diabetes, and autoimmune diseases, as well as heightened emotional and physical reactivity to stressful events. Chronic stress also contributes to significant loss of immune function in the elderly as well as shortening of telomere length and inflammation contributing to cognitive decline. (Seiler et al., 2020).

PTSD Features

According to Sherlin and Nemeroff (2011) the effects of toxic stress can predispose individuals to PTSD and by extension likely complex trauma. Both genetics and transgenerational behavior as well as TBI have their contributions to PTSD as well. PTSD amplifies the stress response, exaggerating the mechanisms we have described to this point. The hippocampus and PFC inhibit HPA activity through CRH neurons in the PVN while the amygdala and aminergic brain stem neurons activate the CRH neurons. The most salient features related to PTSD are abnormal regulation of cortisol and thyroid hormones, and these can be expected to be present in the majority of cases. The dysregulation of thyroid hormones is not as frequently mentioned as is the neuroendocrine dysregulation that is associated with TBI. Thyroid hormones, in particular, are partly regulated by the hypothalamic-pituitary-thyroid axis, which regulates metabolic vs. anabolic and other homeostatic functions through thyroid hormone blood levels. A key marker is very high T3 relative to T4.

Neuropeptide influence comes in the form of the CRH neurons in the PVN of the hypothalamus which integrate information relevant to stress. CRH neurons are found throughout the cortical and subcortical regions that comprise the DMN, in particular, and the locus coeruleus (LC) as well. CRH increases anxiety-like behaviors such as increased startle response as well as fear of new experiences. CRH provides a mechanism by which low cortisol can help sustain stress and fear responses. It is a key mechanism for managing gastrointestinal activity. Neuropeptide Y (NPY) inhibits CRH/NE circuits related to the stress response and reduces NE, so it has anxiolytic properties. Decreases in NPY contribute to noradrenergic hyperactivity. Combat vets without PTSD usually have high levels of NPY.

Hypocortisolism or chronic low levels of cortisol is common in PTSD. It also predisposes people to PTSD by disinhibiting CRH/NE circuits, and reduces hippocampal volume and promotes over-responsiveness. The hippocampus is an important regulator of the immune system, and consequently, this impacts the main region for inhibiting HPA axis. There are also sustained increases in CRH. Low cortisol levels at the time of exposure appears to predispose the individual to PTSD.

High glucocorticoid levels damage the hippocampus, a key component of the DMN network as well. They also reduce immune function by reducing hippocampal regulation. Deficits of glucocorticoids promote activation of the stress response, and this contributes to impaired extinction of stimuli that act as triggers. Of greater psychological relevance, glucocorticoid dysregulation also interferes with the retrieval of traumatic memories.

Dysregulation of the GABA system is another feature resulting in serious sleep impairment and reduction of the main arousal reducing mechanism of the brain, leading to chronic hypervigilance. Chronically elevated norepinephrine (NE) is a major feature that arises in conjunction with decreasing GABA activity. NE neurons project to PFC, amygdala, hippocampus, hypothalamus, periaqueductal grey, and thalamus. CRH and NE interact to increase fear conditioning, encoding of emotional memories, increase arousal and vigilance, and integrate autonomic and endocrine responses to stress. NE release from nerve endings alters blood flow to various organs. NE in PTSD contributes to sustained elevations in heart rate, blood pressure, skin conductance, heightened startle response, and encoding of fear response.

Serotonin projects to amygdala, bed nucleus of the stria terminalis, hippocampus, hypothalamus, and PFC. It contributes to hostility, aggression, and impulsivity and affects appetite, sexual behavior, motor function, and neuroendocrine function. 5-HT availability is frequently low as we have outlined elsewhere and is repurposed in the battle against inflammation. It is also related to behaviors that are like ADHD features including impulsivity, hostility, aggression, depression and suicidality which is partly why ADHD is often misdiagnosed mood disorder. This often results, as explained elsewhere, in chronically elevated alpha that is slowed in nature and appears as elevated theta in the qEEG.

Biofeedback Indicators

During periods of stress, we see progressively elevated EMG activity as skeletal muscles contract to brace for trauma. This is particularly pronounced in men and less so in women who tend to experience contraction of the smooth muscles of the arteries resulting in diminished peripheral vascularity and cold extremities. Electrodermal activity

(EDA) becomes elevated as sweat-gland activity increases during stress, and those who are traumatized experience large variations in EDA activity. This can be measured directly with biofeedback equipment, but I have found the heart rate variability (HRV) more sensitive to emotional shifts, allowing me to see sympathetic versus parasympathetic effects more accurately. Heart rate activity increases from a norm in the 70–85 beat per minute (BPM) range up to the 90–100 BPM or higher as stress increases while at the same time heart rate coherence or HRV diminishes and blood pressure increases by as much as 20–40 points or more. During periods of stress, breath rates shift from an average of 15 breaths per minute (BPM) to 20 or higher, and breathing becomes more rapid and shallow, moving from abdominal oriented breathing to more thoracic chest breathing. Tidal CO2 is diminished as O2 increases with a contradictory shift in cellular blood oxygen levels and shifts in blood gas levels, that is to say, hyperventilation decreases cellular blood oxygen levels.

All of these effects or biometrics can be monitored in real time using biofeedback equipment. During the time when we worked closely with Dr. Jackie Williamson and her team of physicians in Atlanta at the Synapse Clinic in the early 2000s, we compared all the various biofeedback measures against each other and found HRV to be the most singular comprehensive measure of these changes, although breath training had the most impact on all other measures of intervention with respect to biofeedback itself. Hence, we began using HRV exclusively, as it was a very convenient and easy software and sensor arrangement to use in conjunction with the NFB equipment. It provided a nice compliment to the NFB measures as well. It was also easy and inexpensive enough to use so that clients could easily purchase their own devices and practice at home between sessions. When clients did take on this responsibility themselves, clinical outcomes were enhanced.

Chapter 6

Neurophysiology, Electrophysiology and Trauma

The Neurophysiology (and Electrophysiology) of Trauma

One of the key features of stress regarding the brain is excitotoxicity. This insight arises from the research on oxidative stress and its impact on the body as well as the brain (Moritani et al., 2005; Wang & Michaelis, 2010). It is a top-down vector as opposed to the bottom-up vector of microbiome dysregulation and GI inflammation. Environmental and social distress generate both vectors which synergistically produce a bidirectional response in the body.

Chronic levels of stress produce elevated set points in the reticular activation system (RAS), and this keeps norepinephrine levels high. This also generates excessive activity both cortically and subcortically as many, if not most, people respond with worry and rumination, activities that utilize excessive glycogen and other energy resources. Perceptual processing is chronically elevated, and this increases bi-hemispheric coherence in the beta frequencies 15–20 Hz. When this occurs in the association cortex, there is increased hypervigilance and sensitivity to potentially dangerous stimuli in the environment. This, in turn, degrades into perceptual filtering problems as the system becomes exhausted.

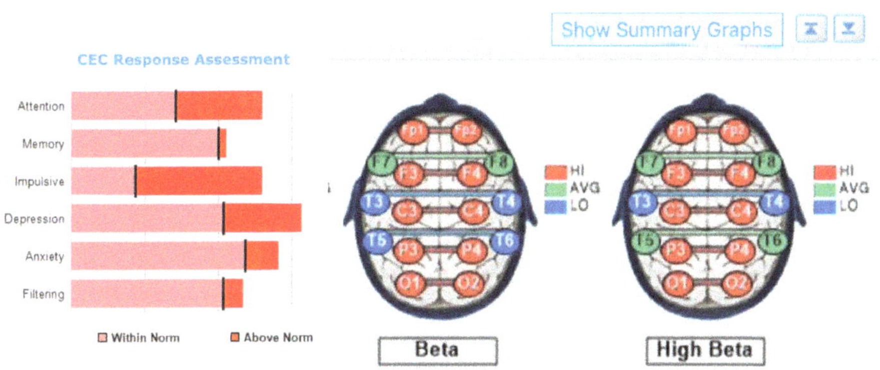

The visual processing cortex in Brodmann areas 18 and 19 also becomes chronically hyperactivated, overdriving amygdala inputs. People become more sensitive to light and sound as well as proprioceptive stimuli. This affects food texture experience as well as fabric texture experience, and many clients report problems with eating and clothing selection to manage these chronic and distracting issues.

Hypervigilance also increases magnitude in the beta frequencies, especially those related to excess cognitive processing in the 15–20 Hz range (Heller et al., 1997; Meehan & Bressler, 2012). This excess processing is related to anticipatory worry, reactive anxiety or panic, and rumination (Engels et al., 2007). Unfortunately, excess EMG is also highly associated with hypervigilance and masks the distribution of excess beta with subharmonics in the hi beta range infiltrating the beta filters. Consequently, it is difficult to ascertain beta magnitudes on the typical map unless it has no significantly high amplitude hi beta. The map below has lower than normal hi beta indicating it is artifact free, and we can see the real beta is elevated in the posterior region when it is adjusted for amplitude.

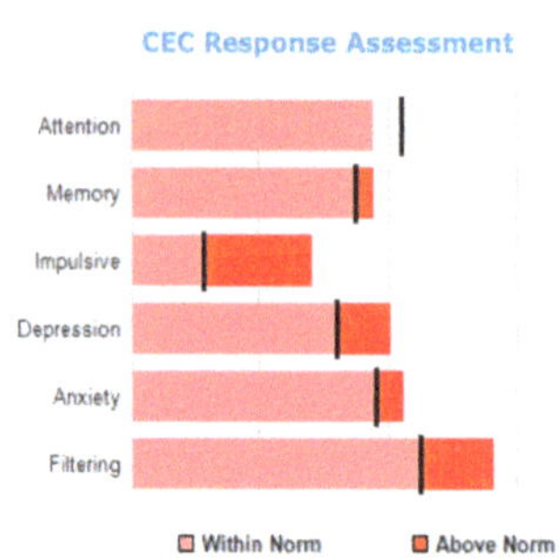

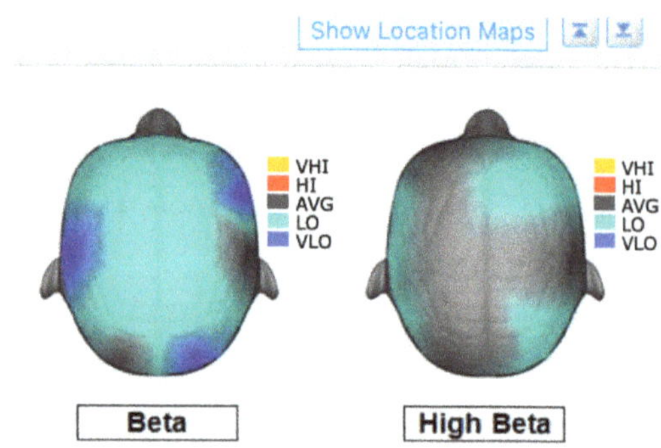

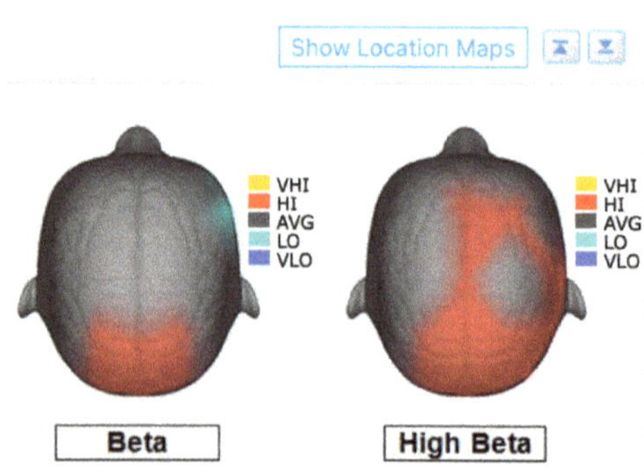

Traumatized individuals continually review the past events related to their trauma in the form of chronic rumination with related shame-based emotional constellations (Bradshaw, 1988; Forbes et al., 2020). At the same time, they psychologically brace for future events of a similar nature with repetitive anticipatory speculation and worry, grounded in the frontal networks, especially the prefrontal BA area 10. At the same time, the heightened RAS overdrives the right hemisphere amygdala and global attention, including heightened body language decoding and facial decoding as well as heightened decoding of tonal inflection information, issuing from the right posterior parietal and temporal parietal junction respectively. This results in overactive right frontal attentional networks and their global gestalt processing for danger, which eventually hands off the localization task to the left hemisphere and associated excessive anticipatory worry (McGilchrist, 2009). In addition, there are also beta increases in the right hemisphere relative to the left hemisphere with the increase in arousal and anxiety.

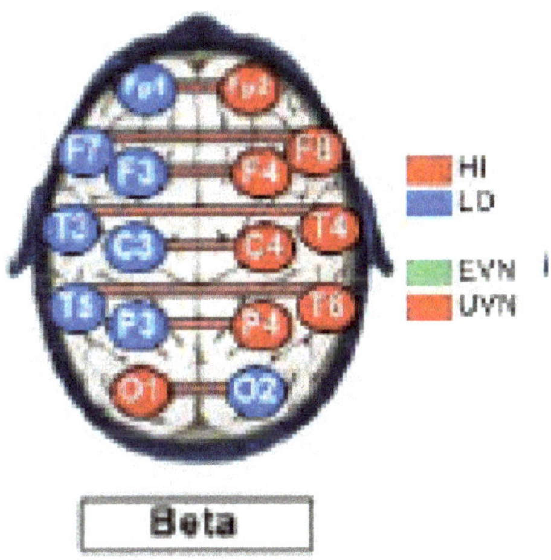

The red locations on the right side of the image indicate beta amplitude on the right is higher in amplitude than its contralateral equivalent, most of which are blue indicating they are lower than the right side.

This is likely a standard systems-based response evolved over millions of years, with each network and response establishing itself over the phylogenetic history of the mammalian species. The overall effect of this process is to chronically overactivate the physiology as described in the previous chapter, generating adrenal, cortisol glucocorticoid elevations that overstimulate and generate inflammation in the GI tract. This bottom-up activity precipitates an inflammatory response in the brain, reducing neuronal resources and gradually depleting high frequency activity in the beta range, and generating more idling in the thalamocortical system, with elevated alpha resulting.

At the same time, the overactivated cortex consumes very high levels of glycogen and related metabolites. This parallels the changes in GI activity and zonulin levels described in chapter 5. A great deal of glutamate is generated from the heightened metabolic activity (Wang & Michaelis, 2010). It is the task of the supporting microglia to convert this back into glutamine for recycling in neuronal metabolism. However, with heightened arousal and hypervigilance, sleep becomes contemporaneously problematic and diminished for most individuals (Walker, 2017). The usual nightly cleansing of these excess metabolites through the action of the brain's lymph system is diminished, and the glutamate then builds to toxic levels over time, driving microglia into apoptosis and death. This generates the pro-inflammatory response at the same time the inflammation in the GI precipitates the pro-inflammatory response through the kynurenic pathways, as discussed in chapter 5. This dual inflammatory effect drives calcium ions through the NMDA channels of the neurons, which mitochondria attempt to absorb to protect the neuron (Moritani, 2005). This, in turn, burdens the mitochondria, and less ATP is produced. This reduction in ATP undermines the sodium and potassium pumps of the neuron, which provide the electrical charge across the cell membrane. This interferes with the neuronal interstitial current as well as collective pre and post synaptic EEG activity, i.e., brainwaves. The net effect is a loss of high frequency activity which animates the cortical process. Over time, the amplitudes in the beta range diminish while the brain shifts more progressively into a routinized stand-by mode, dominated by alpha, as the cortex does not have enough energy to sustain active creative worry and rumination.

This domination of routinized activity results in repetitive metastable network patterns that repeat indefinitely. In terms of the frontal executive networks, LeDoux (1996; 2016) notes that this has the net effect of gating input from the right amygdala. So, it is likely anticipatory worry has a stabilizing function that mitigates excessive chronic arousal from a chronically overstimulated amygdala. The major impact on the brain is in the ventral tegmental area and the locus coeruleus which activates dopamine and norepinephrine pathways in the brain, increasing arousal and vigilance. This does not, however, block the other six or more major amygdala outputs to the rest of the physiology, especially the heart, the gut, and the adrenals along with the other physiological events discussed in the previous section. The RAS remains locked in to a higher arousal level as well. Inputs to the hypothalamus accentuate sympathetic activation and can induce tachycardia, GSR and sweating, pupillary dilation, and increased blood pressure. The parabrachial nucleus in the nucleus solitarius increases respiration and shifts the breathing to thoracic rapid shallow breathing. The brainstem reticular formation increases reflex speed, and in PTSD, results in elevated startle response. The periaqueductal gray matter can, depending on circumstance, induce freezing, while the trigeminal and facial nerve nuclei mobilize the automatic familiar face of fear. The paraventricular nuclei of the hypothalamus releases ACTH and consequently stress hormones.

As the brain loses sleep and also loses sustenance due to the microbiome dysregulation; as inflammation increases, compromising ATP output; and as the beta amplitudes diminish with the loss of energy to produce high frequency activity, the brain seems to slip into an evolved secondary defense mechanism. The routinized activity of worry and rumination results in elevated alpha activity as beta wains. The motor system neurons that drive the cognitive macros of worry and rumination through the basil ganglia and the cerebellum ramp up their activity (Schwartz & Begley, 2002). Much as a juggler can talk and juggle at the same time, the brain juggles its macros and continues to carry on other functions, but at a diminished level. In particular, we see elevated posterior alpha as the posterior cingulate locks into this repetitive process. Posner and Raichel (1997) and Peterson and Posner (2012) noted the cingulate's tendency to operate this way when it became stuck on a particular stimulus. The cognitive correlate of this is perseveration and repetition. For some, with the genetic predisposition, this morphs into OCD. Eventually, the anterior cingulate becomes less active in attaching itself to salient stimuli, because depressive features enter the picture and there is a loss of salience and interest for the individual.

Depression not only brings heightened inflammation that reduces serotonin levels, but it is also associated with reduced dopamine (Belujon & Grace, 2017), and other neurotransmitters related to hormonal sources such as estrogen and testosterone (Purves-Tyson et al., 2014). Deficits in the enzyme methylenetetrahydrofolate reductase

(MTHFR) contribute to failure in the folate cycle that leads to disruptions in the BH4 cycle for synthesis of the major neurotransmitters of serotonin, dopamine, and norepinephrine. Although the details are not fully understood, the changes of EEG asymmetry clearly follow (Soutar, 2017a). Alpha increases in amplitude in both anterior and posterior regions as brain idling increases, and alpha begins to dominate the left instead of the right hemisphere. Withdrawal behavior begins to dominate with a loss of interest in usual activities (Davidson, 2000; Solomon et al., 2014). This is a key transition point between anxiety and depression as outlined in my article on oxidative stress and the qEEG (Soutar et al., 2017a). A further paper demonstrates how most individuals begin reporting depressive symptoms as anxiety reaches a moderate score of 25 on the ISI and beta asymmetry approaches 90%. Those who do not shift into a depressive state continue to score higher above 25 on the ISI and exhibit panic disorder features (Soutar, 2021).

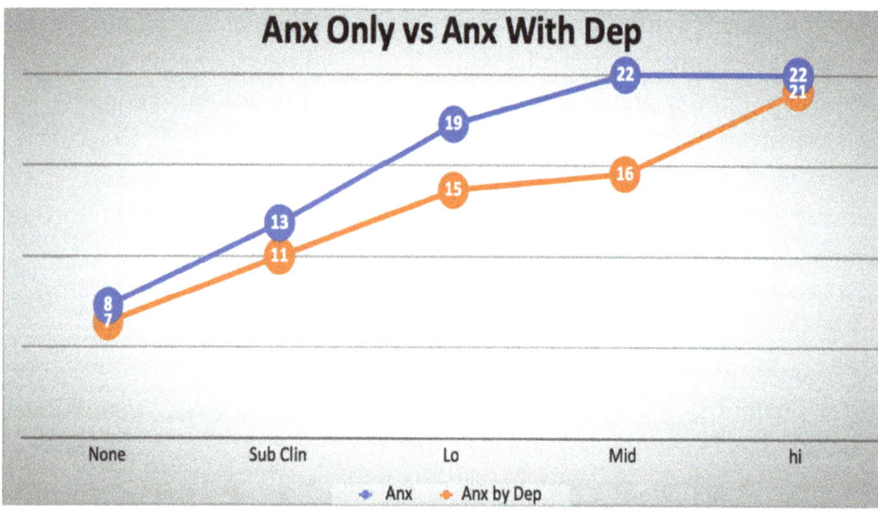

In the chart above (N = 2393), the blue line shows mean anxiety scores of those reporting subclinical or no depression. The orange line shows mean anxiety scores in those with depression as it increases. Note that the anxiety scores are lower in those with depression, suggesting that depression modifies or reduces the experience of anxiety. Anxiety scores never reach above 25 without depression being present. In those with depression, anxiety scores do reach as high as 30, but this group comprises only .01% of the entire sample, so it is very rare. 137 individuals reach levels of high depression in this group which is .06% and so significantly greater.

Their cortisol system appears to fail dramatically to control progressive elevations of the RAS set point and stress hormones. As the individual develops depression and growing inflammation, the EEG begins to slow in dominant frequency. Those individuals with the greatest levels of physical health problems, above 250 on the New Mind Physiological Checklist, and who are also depressed, demonstrate the greatest level of EEG slowing, especially in the alpha frequency range. A score as high as 250 occurs in only about 10% of the population.

It is likely that for those demonstrating the higher scores in depression measures, sleep deprivation contributes to this slowing, as well as the appearance of elevated delta in the eyes open state, along with especially elevated alpha. In the example below, thc cyes closed map has a normal level of delta with the eyes closed (except for the temporal lobes), but the eyes open map reveals a high level of global delta. This is typical of sleep disorder or insomnia. Notice the appearance of diffuse elevated alpha, which indicates brain fog.

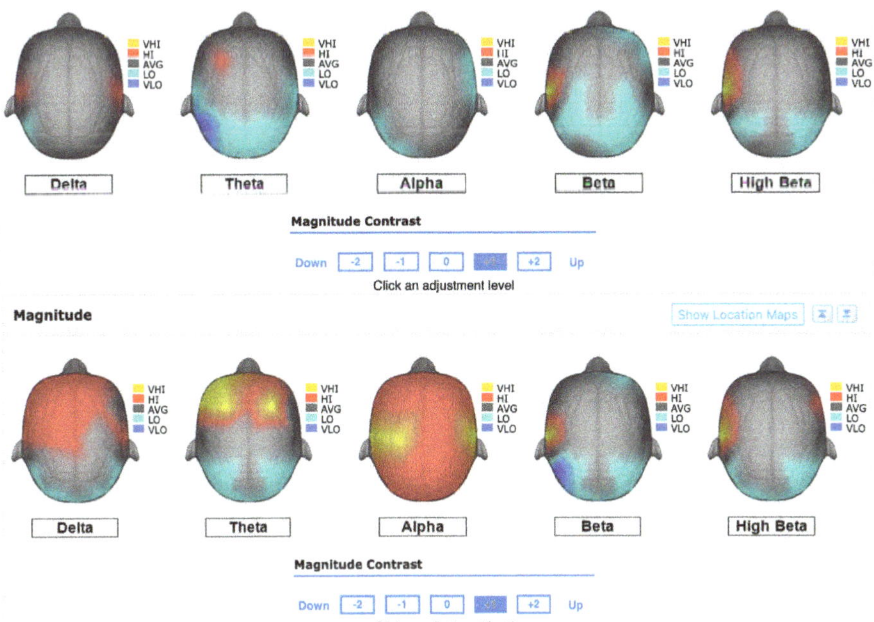

Sometimes this alpha is so slow that it is interpreted as theta by the digital filters and can be seen as very low frequency alpha combined with very fast theta as above. Theta varies very little in dominant frequency in the normal population. Often elevated theta is really slowed alpha, so slow it is picked up by the theta filters.

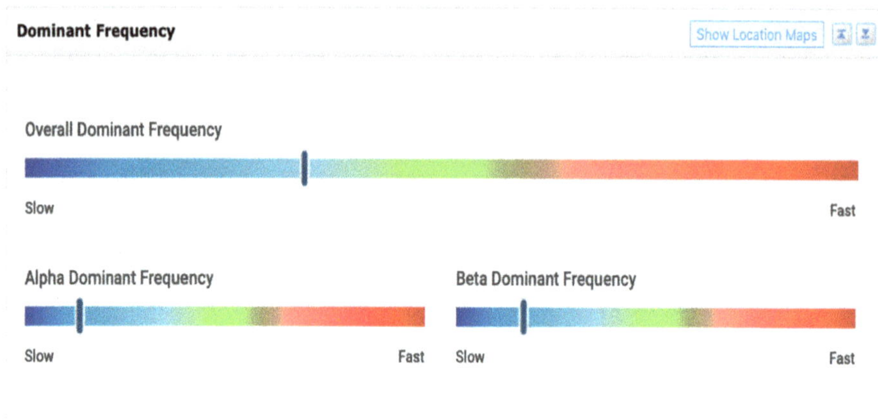

In the last decade, this has led to the identification of many features attributed to ADD, such as elevated frontal theta, as being actually due to a sleep disorder and mood issues (Arns et al., 2013; Núñez-Jaramillo et al., 2021). This forces us to consider that many attentional problems are due to anxiety and depression, issuing from excitotoxicity related to familial distress, arising from dysfunctional transgenerational behavior and associated trauma. All of this is consistent with the ACE study reviewed in chapter 2 where toxic stress is cited as the culprit, but whose definition agrees nicely with complex trauma.

Other Dissociative Patterns in the EEG

Besides mind chatter associated with low or significantly elevated alpha, there appear to be additional strategies that the brain adopts to cope with trauma. All these strategies enable the dissociative process arising from and generating the disconnect between the cortex and the limbic structures. Over the last two decades of training and collecting data at NewMind Center, I have identified several major distinct EEG amplitude patterns of component bands in the session trend screens. Although they also involve other unique aspects, reflected in the other neurometric dimensions, the trend amplitude characteristics which we read as magnitude in the qEEGs are the leading identifiers.

Theta is a key frequency related to memory and emotion, generated from the septal hippocampal oscillatory network (Buzsaki, 2006). It is this same network that helps to bind cortical networks through phase coupling and nesting dynamics involving gamma (Meehan & Bressler, 2012). Gamma is really an intermediary (Schroeder & Lakatos, 2009) binding limbic and cortical activity, and this appears to be lost upon many in the field. A key region of the brain for this process with respect to emotion is the conversion zone of the temporal network involving the insula, the salience network feeding into the medial prefrontal (LeDoux, 2015; Yankouskaya et al., 2022; Bolton et al., 2020).

During heightened emotional processing, global network activity increases dramatically, especially in the self-referential DMN during very pleasant emotional images (Sambuco et al., 2022). Theta has been demonstrated to be associated with mind wandering and imagination (van Son et al., 2019). This follows from the fact that theta has direct inputs from the limbic oscillators into the posterior theta circuit of the DMN, and the DMN has dense fiber pathways to the limbic structures (Zheng et al., 2021). It has long been known that theta is highly associated with memory and emotion. It also elevates significantly globally during fantasy and daydreaming. This can be readily seen in dynamic terms while observing the trend screen as clients dissociate into a mental image as they train. I have asked clients to imagine different scenes and monitored their theta. I found those most practiced at imagination can generate very high levels of theta during imaginary episodes, especially in the DMN at Pz. I believe that this is the reason that Pz is such an effective location for AT training and why it is associated with strong emotional memories. Bill Scott has indicated to me that he came to similar conclusions (personal communication).

Conversely, some developmental traumas result in a complete suppression of emotion and significant theta deficits in magnitude as well as theta hypocoherence in the temporal lobes and insula region, indicating poor emotional regulation. Referring to the same literature explaining the mechanisms of high theta in the previous section, we can reason that low theta would be an absence of strong emotion and imagery due to absence of input. Low theta is indeed a common condition of stress and anxiety upon inspection. A NewMind archival statistical analysis N = 1253 adults reveals that there is a very high correlation between low theta and anxiety, as well as poor memory. This pattern is especially strong in the posterior at Pz. The ISI anxiety scores in the normal range of < 10 have significantly higher theta than those with clinically significant scores > 20.

t-Test: Two-Sample Assuming Unequal Variances

	Theta Anx scores > 20	*Theta Anx Scores < 10*
Mean	8.054377682	9.023960784
Variance	8.431550581	17.85472816
Observations	233	1020
Hypothesized Mean Differ	0	
df	485	
t Stat	-4.184389233	
P(T<=t) one-tail	1.69748E-05	
t Critical one-tail	1.648001465	
P(T<=t) two-tail	3.39497E-05	
t Critical two-tail	1.964867287	

My clients with complex trauma who have very low theta find it nearly impossible to uptrain in many cases, and I believe this is the reason Green & Green (1977) found this a common issue. Many clients with highly levels of anxiety present this way, especially men with strong EMG readings and often low voltage/power qEEGs as well.

Finally, a common pattern in trauma is the EMG factor which has been previously discussed. EMG, usually represented as significantly elevated hi beta amplitudes or magnitudes greater than 8 μv, can be observed in a great many clients. This component band can be seen to move up and down dynamically as clients encounter various triggers while watching videos during NFB training.

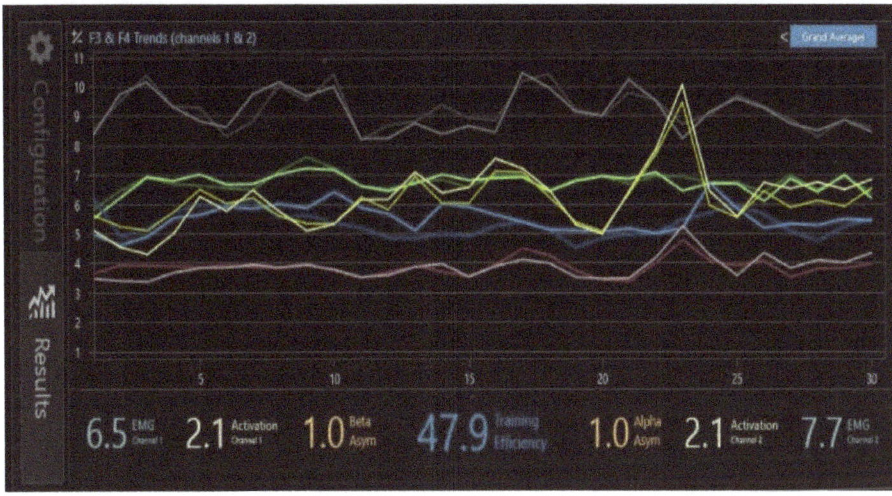

Around 24 minutes the spike in beta indicates the client is bracing and experiencing feelings related to trauma, and then adapting and perhaps integrating them.

As can be seen from the foregoing, trauma results in a variety of coping methods that generate a subtype of features identifiable in the EEG neurometric dimensions and component bands. Trauma has many faces. I will explain in more detail how these identifiers are operationalized in chapter 8 wherein we discuss the various trend screens in AT training and their significance as well as their management.

Brain Networks and Trauma

Much of what is presently known about brain networks and trauma is speculative. One prominent investigator, Olaf Sporns, commented about a decade ago that our present status in the neurosciences regarding understanding brain networks is equivalent to what Columbus knew about the world in the 1500s (Meehan and Bressler, 2012). Not much has changed. It is tempting to speculate, and overreach based on what we actually know, especially when desperation to help those suffering is combined with marketplace profits. Much of the science behind the use of SSRIs for depression based on research has proven invalid (Ioannidis, 2005; Whitaker, 2010). A recent study by Khan & Brown (2015) found that most antidepressants had at best a 13% advantage over placebo. We now understand that much of the mechanism of action of this class of drug is more due to its action on inflammation rather than synaptic increases in serotonin. Based in general on what we know about the long-term negative impact on mental health of supposedly double-blind tested drugs it is not unwarranted to describe their implementation as highly experimental. We presently understand that their deployment was not based on sufficient evidence. At the time, most believed the risk was warranted. NFB poses a much smaller risk than drugs and surgery, but we should still speculate and implement protocols with careful consideration of the limitations of our knowledge. The NewMind symptom trackers regularly show reductions in client reported levels of symptomology as high as 40 to 60% or even higher due to NFB protocols. This is average over hundreds of clinics over a 15-year period. It is consistent with pre-post measures of the effects of AT in prior investigations using well-constructed assessment tools such as the MMPI and Beck Inventories. In this spirit of speculation, we embark on a brief review of present considerations in the literature regarding the activity of networks with respect to trauma.

Menon (2011) presents a picture of large-scale networks, the triple network model of psychopathology, that has continued to be well supported in most investigations since his publication. This includes the latest finding by Nicholson et al. (2020; 2022), which we will consider shortly. To best understand the concepts, it is first critical to review at a basic level the fundamental components of the arguments proffered. We begin with the celebrated insight of Raichle that what was thought to be the background noise of the MRI bold signal was the idling of the primary networks of the brain (Raichle et al., 2001). It was a little like finding out the annoying background radiation of the universe was from the Big Bang. After a flurry of investigations Buckner published a jaw dropping paper outlining what is now known as the default mode network in 2009 (Buckner, 2008). This network is the centerpiece of the central networks of the brain, known as large scale networks, that engage in self-referential activity. This includes interpretation of mirror neuron activity that reports the possible meaning of the body language of others and theories regarding what they might be thinking or theory of mind as we call it in developmental psychology. It involves moral decision making, autobiographical memory, and envisioning the future as well.

The center hub of the DMN is the precuneus which is situated in the posterior region of the brain and is included in the area around Pz which is a preferred placement area for AT training sensors. This network includes temporal lobe connections and midline connections to the ventromedial prefrontal portions of the brain, including attentional hubs in the anterior cingulate. This hub has major connections to the limbic structures as well, especially through theta rich anterior fiber pathways (Zheng et al., 2021). The limbic structures are major sources of emotional and memory inputs.

The prefrontal cortex (PFC) of DMN-PFC includes ACC and the medial frontal region. The PFC controls the amygdala cortical inputs and inhibitory control over stress response. The amygdala, which is part of the subcortical DMN, is especially too sensitive to faces, stressful words, cues, and trauma reminders, and is a key trigger source. It plays a key role in extinction of fear response and integration (Sherin & Nemeroff, 2011). PTSD results in reduced volume in this area of the brain and reduced inhibitory control. Decreased activation of medial PFC appears to be common in PTSD. It is interesting to note that SSRIs restore mPFC activation!

Another major large-scale network of related importance is the central executive network (CEN). The DMN and the central executive are part of a major set of networks that dominate processing and direct behavior (Menon, 2015a; 2015b). The center of this network's activity is around the frontal lobes, in contrast to the importance of posterior processing in the DMN. The CEN network also has strong intrahemispheric connectivity to the parietal regions that flank the DMN. This network is central to executive functioning and decision making in the brain.

A third network of related importance is the salience network (SN). This network is anchored in the major neural convergence zones of the temporal lobes, and is heavily linked to limbic structures as well. It is also heavily linked to the anterior cingulate and the orbital frontal region which is the key executive region of the brain, Brodmann area 10. This network shifts the center of attentional activities in the brain from the anterior external attention system involving the central executive functions to the posterior DMN system which involves mentalization and self-referential processing. Much of this shifting is based on emotional valancing arising from interaction with the social environment.

These major large-scale networks shape much of our conscious activity from the interaction of bottom up (emotional response) and top down (executive perceptual input) that engage or disengage the DMN at various levels of activity. Cortical and subcortical processing centers related to autonomic activity overlap and support these three major network functions through highly complex feedback loops. The autonomic response, discussed at length in the earlier part of this chapter, delivers much of its input through the ascending reticular system, driving thalamocortical inputs, and the ascending vagal system, driving the insula, which has strong inputs to these networks as well. When external stimulus events of high salience, i.e. threats, trigger the amygdala's early warning system, these ascending systems activate the central executive and external attention system and accelerate executive processing.

The central executive network, at a certain point, becomes overdriven in both the perceptual association cortex in the parietal region as well as the frontal attention networks in the orbital frontal region. At the same time, the DMN is also hyperactivated by limbic inputs and the salience network shifts into excessive self-referential processing, but these events occur in complex stages. Depending on circumstance, either the CEN or the DMN become the more dominant center of gravity in processing terms. At high levels of triggering both are shut down as autonomic activation overdrives both systems and individuals enter a state of panic. Various automated responses to danger then override the entire organism into a fight or flight scenario.

One of the key insights to trauma emerging from brain network analysis is this salience network-based model of psychopathology. The degree of engagement or disengagement of the DMN with the salience network results in significant baseline alterations in self-referential activity.

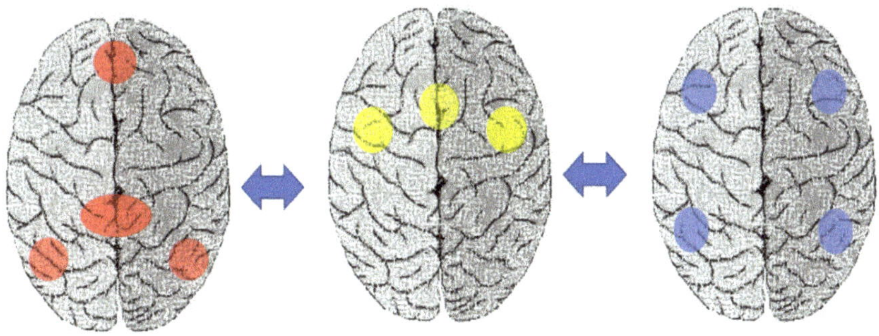

Triple network model: The salience network switches the attentional resources between the interior oriented self-referencing processing of the default mode network to the external oriented central executive network. In disordered individuals, excessive rumination and slowing in the EEG draws the emphasis on the internal processing.

Excessive rumination and obsessive thinking, driven by emotional and autonomic inputs, arise in concert with withdrawal related thinking and chronic overactivation of the DMN. This may involve increased posterior beta activity which is often overshadowed by heightened EMG response in the scalp muscles. This in turn overdrives the amygdala and autonomic functions in a negative bi-directional loop, hence the increased baseline arousal of complex trauma. The brain may respond with increasing posterior alpha activity such as mind chatter to block inputs to manage arousal more effectively at a functional level suited for any required social interaction. It may also resort to one or all of the other strategies discussed in the trend screen profiles in trauma previously reviewed.

Chronic dissociative processes related to withdrawal behavior and alpha asymmetry pull the client into excessive self-referencing activity without sufficient central executive based reality testing through social interaction with others. Over time clients grow a network profile that captures their salience network in aberrant intrinsic organization and interconnectivity with the DMN and CEN (Menon 2015a; 2015b) as displayed below and keeps them focused on internal automatic emotional and automatic thinking processes, especially negative rumination.

This triple network model has led many neuroscientists interest in PTSD and complex trauma to focus on identifying the key structures and network patterns in more detail that might identify trauma as well as lead to clues on how to mitigate its affects with NFB.

Terpou et al. (2020) looked at the periaqueductal grey region (PAG), the medial prefrontal region, the precuneus and the DMN and found that individuals with PTSD in an exposure state were very driven from the bottom-up stimulation of the PAG in these areas and other regions of interest as well. These were important findings. However, they appear to overlook and fail to discuss well-established findings as related by LeDoux (1996; 2015) that the amygdala drives the hypothalamus which in turn drives the PAG, all of which are linked in a "serial defense network" which explains the bottom-up stimulation better than their speculations. Exposed individuals have very high and rapidly escalating input from the amygdala.

Drilling Down Further

Nicholson et al.'s research (2020; 2022), for those who are interested in going deeper, reviews the EEG and fMRI literature in great depth and with excellent economy. The exact details of the hubs and nodes of the DMN and how they interact are to date still being worked out. To understand how the implications of these new insights regarding the DMN apply to NFB and AT training, it is of value to define the DMN in some more detail.

The DMN has two centers of gravity in terms of processing, the anterior portion, which is in the frontal lobes, and the posterior portion, in the parietal lobes. Specifically, the DMN with the affiliated 10-20 locations (in parentheses) consists of the medial prefrontal cortex (MPFC; Fpz), the posterior cingulate cortex (PCC; below Cz and into Pz), precuneus (below Pz), inferior parietal lobe (IPL; P3-P4), and the lateral temporal cortex (LTC). In most presentations, this temporal lobe designation is often poorly defined. This key temporal part of the DMN network is BA21 of the medial temporal lobe, and it links up to the subcortical regions involving the hippocampal formation (HCF). So the core regions of the DMN are the MPFC and PCC. Newer more accurate imaging includes the dorsal prefrontal cortex, and other subcortical regions including the thalamus, the septal nucleus, and the nucleus accumbens

(Alves et al., 2019). I add these because they are noted in the NFB literature reviewed below as separate from the DMN.

The self-referential DMN has been described in some detail up to this point, but it has become a focus of interest in neuroimaging research on PTSD as well as the qEEG research on PTSD. The qEEG research combined with fMRI has lead to an identified pattern of not only anatomical distribution related to PTSD but also to functional patterns as reflected in the EEG.

At the moment, only one functional pattern and a limited set of protocol solutions have been acknowledged, but our clinical database, involving over 500 clinics, suggests that others will be coming. AT training is an underacknowledged protocol in this community of researchers involving Nicholson, van der Kolk and their associates because they have not extensively utilized this protocol and are seeking rapid immediate solutions like their drug protocols which bias as well as distort their expectations.

Understanding the DMN and how it responds to NFB is of great value in understanding the underlying mechanisms of action in AT training, and helps to explain its efficacy as a protocol. Re-experiencing and retrieval of autobiographical memories related to trauma, often referred to as triggering and exposure, activates the DMN in experimental subjects differently than in the control subjects (Thome et al., 2019; Fenster et al., 2018). The general difference between subjects without PTSD and subjects with PTSD (as well as complex trauma) is in the relationship between the anterior and posterior components of the DMN. Nicholson et al. (2022) have done an excellent job of laying this out, and most of what follows is from the neuroimaging literature review as well as the qEEG literature review of their investigations. This includes especially their own most recent research findings. This team, in one form or another, has been working with fMRI and qEEG together on this specific topic for well over a decade and a half.

Both anterior and posterior portions of the DMN display disrupted connectivity, as defined in the fMRI literature, both at rest and during executive tasks in individuals with PTSD. This disruption is between the anterior and posterior region, but there are also connectivity issues within each of these regions.

The increased activity within the posterior DMN involves increased connectivity between the posterior cingulate cortex (PCC) and the precuneus which is just posterior to it. The PCC is also reportedly overactive in terms of BOLD signal. In our clinic, it is not uncommon to see elevated posterior alpha, and often posterior beta, in individuals who are perseverant. This beta as mentioned, was part of Engels et al. (2007) findings regarding anxiety involving the precuneus and inferior parietal cortex, especially in the right hemisphere. This occurs as significantly increased beta at Pz, P3, and P4 which are often picking up the inferior parietal cortical regions. Going back several decades Posner and Raichle (1997) noted this region was critical to the attentional component process of letting go of an object of interest. Earlier postmortem studies had also found this region damaged in individuals with perseveration issues (Crosson. 1992). CBT therapy has also been reported to reduce PCC activity.

It is also interesting to note in reference to this, that the majority of the qEEGs we have done using sLORETA over the years have shown excessive beta activity in the posterior cingulate region (PCC) of individuals with trauma. At first, I thought it was a problem with LORETA, only to discover it was not when reviewing the first publications on the DMN by Buckner et al., (2008). When down-training this beta activity we found EEG patterns shifting into a more normative pattern with both our equipment during training as well as in pre-post mapping using several different databases available at the time.

In contrast, for those with PTSD, within the anterior component of the DMN we find connectivity decreases within the medial prefrontal cortex (mPFC) and specifically the vmPFC which is more related to emotional processing and episodic memory. Decreases of volume and activity in these regions are common in PTSD (Sherin & Nemeroff, 2011). In comparison, in terms of regions, the dmPFC activates during self-related emotional processing during exposure. It plays a key role in networks related to processing. Interestingly, the more the dlPFC is activated, which is more related to updating and tracking errors in prior beliefs (Bzdok et al., 2013) and recently recognized as part of the DMN, the less PTSD symptoms are present. The same occurs when the anterior cingulate cortex (ACC) activation increases. These areas are key components of the CEN. When the CEN normalizes in activity, emotional regulation improves along with reduced DMN and SN activity. This is consistent with what Menon's review of the literature found (Menon, 2015a). DMN connectivity is increased more than the normal CEN and salience connectivity that occurs during executive processing in individuals with trauma. The outcome is suboptimal downregulation for DMN that indicates difficulties disengaging from DMN.

What all of this points to is stabilization of dissociative shifts to the self-referencing system by the SN and concordant reductions of DMN self-referencing activity. The pull of attentional gravity is toward the CEN and the external attention system. It is the opposite of what occurs in the dissociative process that diverts attention excessively to the DMN activities. The more the dlPFC activates, the less intrusions of negative automatic thoughts and emotions with concomitant arousal problems. One study found that when the amygdala region was down-trained using fMRI, there was increased connectivity between the amygdala and the dlPCF. This may be part of error integration, updating and the integration process. This shift also increases resting state connectivity with the precuneus which both participates in the CEN and may indicate error detection and integration feedback to the DMN self-referencing process. CEN increases are shown to improve emotional self-regulation. This is consistent with LeDoux's findings that the ACC, OFC, and these regions need to increase connectivity in order to retrieve implicit memories and port them into cortical narratives, i.e. trauma integration (LeDoux, 2006; 2015).

The left anterior insula became more active in exposure states along with the right cerebellar activity. The insula activity is tied to increased SN activity related to increased hypervigilance, hyperarousal, and re-experiencing. The cerebellar activity was correlated with more intrusions and negative alterations in cognition, which would actually be more negative automatic thoughts relying on cerebellar timing and control functions. The amygdala and PCC connectivity increased with PTSD symptoms, and also decreased with reduced PTSD symptomology, suggesting limbic structures overdriving the posterior regions of the brain. Failure of frontal activation, as in worry (as we have seen in LeDoux's analysis), also makes the individual more vulnerable to overwhelming fear.

Of further interest was that hippocampal regions with deep connections to the medial temporal regions of the DMN increase in activity and connectivity with greater DMN activation and connectivity during exposure states. This indicates increased memory activity as greater activation in fMRI means a higher level of septal hippocampal oscillator and theta EEG production.

Gruzelier and Egner and The Role of Theta

Germaine to this topic of theta production is the theoretical perspective presented by Egner et al. (2002, 2003) where he experimentally isolated the effects of theta training without the usual confounds in past research. There was at the time concern that theta could even be enhanced through NFB and what role alpha had in the process in spite of the Greens' (1977) findings. Egner set out to determine empirically what effect this unique combination of training alpha and theta had on the EEG activity of the brain. His findings appear to indicate that this method of theta enhancement does actually increase theta production during sessions, as sessions progressed. So, subjects were more significantly enhancing theta activity in each successive session. It appears that the main effect of AT training is enhancing theta activity. So, what was the role of theta frequencies in the brain that resulted in the outcomes of AT training?

Gruzelier (2009), a major researcher in AT, took pains to consider the general mechanisms of action concerning how increasing theta resulted in the outcomes they observed. He reports that he began using AT in 2003 with musical performance in isolation from other interventions and compared it to beta and SMR training. This was due to Egner's findings and his own concerns about confounding effects. He wanted to isolate the effect of AT as much as possible. He found that only AT was correlated with improved musical performance. They also found creativity enhanced. Anxiety reduction, on the other hand, was improved in other NFB interventions and associated techniques. In further experiments the effect of theta was deemed independent of anxiety reduction and mostly found to improve performance through improved creativity. This finding is consistent with the Greens.

Interestingly, he takes the significance of this one step further and considers the role of theta at the network level. He observes that theta has wide ranging correlates with many behavioral and neuronal processes. He notes that theta is associated with cognitive processes such as working memory, recognition memory, mental calculation, learning, spatial navigation, conscious awareness, regulation of action, and sustained attention. He comments that animal studies show that when orienting to novelty theta increased, and at root it is related to long term potentiation (Buzsaki, 2006). This consistent with studies we have cited regarding anterior midline theta in humans, and the role of the SN in directing the external attention system and its function within the CEN. He notes that it also is consistent with the findings of two thetas research of Kirk and Mckay (2003) regarding the two circuits that manage memory and emotion. He reminds us that throughout the classical Papez circuit, theta plays a major role in emotion and motivation as well as the amygdala. So, arousal networks, memory networks and emotional networks are all involved.

To his credit, before Menon's triple network model, he observes theta's role in coupling and integration of widely distributed networks, and that theta was part of top down regulation of these widely distributed networks. This is related to making novel cognitive associations through integration of these networks. This involves more than a cognitive enhancement process. His research on music and dance indicates theta training affected qualitative dimensions such as confidence, commitment, emotional expression, enjoyment, and influenced deportment, breathing, diction, and stage presence.

In summary then, from this perspective, AT training is producing low frequency changes in fundamental networks relating to all of the above functions, which contributes to openness, to novelty and new solutions, as well as the creativity to operationalize them at a very fundamental neurological level. These are exactly the key skills and disposition required to reconsider past memories, reframe them, integrate them and draw on them to enhance future interactions with confidence and improved outcomes. I will draw upon these arguments more and connect them to a psychological theory regarding AT and trauma in a later chapter.

Van der Kolk NFB Research

Van der Kolk (2014) speculated, based on the research to date, that an inhibited medial prefrontal region, as discussed above, is likely to result in problems with having a sense of purpose and direction, as well as the ability to make decisions and plan action. He found this deficit consistently a part of the complex trauma features, and possibly undervalued in its contribution to the condition. In addition, he reasoned that in an effort to shut off terrifying sensations, clients deadened their capacity to feel fully alive. This is likely another aspect of the emotional numbness that many with trauma report, related to the loss of emotional valancing that often arises from intense anxiety. Van der Kolk suggests that this profoundly affects self-awareness, and is consistent with the findings of Damasio in The Feeling of What Happens (1999), which apparently influenced his thinking in this area. We tend to overlook the importance of self-awareness in all this research, and how much it is related to shifts in anxiety and depression. Damasio emphasizes that emotion is a key feature of self-awareness and consciousness itself, so that blocking emotion is likely to also reduce both self-awareness and consciousness. Damasio said that as a neurologist, the only way he could often distinguish between a conscious and unconscious patient was the level of emotion in their presentation.

Van der Kolk's recent entry into research on NFB efficacy and trauma was predicated partly on his association with Ros and with Lanius as well as with other more recent studies (Gapen et al., 2016). His introduction to NFB was apparently far less influenced by the protocols of Green, Peniston, and Scott. The Gapen team, which included van der Kolk, argued that newer imaging studies suggested that calming the right hemisphere amygdala was a superior strategy to AT, as it was more direct. Apparently, they were unaware of the research indicating that the locations of the brain at O1-O2, where the first AT research focused, also had connections to the amygdala, involving an important visual fear pathway (Silverstein & Ingvar, 2015). His training strategy also tends to reflect older theories of NFB by focusing on ROIs and a more modular model of the brain. More recent studies make it clear that the brain is so interconnected that it has to be considered as a system. As such, there are likely many areas of the system where inputs can affect equal or greater normative reorganization depending on individual differences. In my opinion, their failure to find T4-P4 training superior to T3-T4 training in terms of statistical significance actually reveals their unfounded and theoretically based bias in favor of ROI types of training.

A 2020 follow up study to Gapen's (Rogel et al., 2020), also affiliated with van der Kolk, was more rigorous in design with a greater N =

37 instead of 34, and also random assignment to an experimental group or wait list. Subjects were also required to continue therapy and medications in both groups. It did, however, use the same protocols and use the same training. Their outcomes in this design were apparently superior as well. They used 24 sessions instead of the Gapen 40 sessions, and this produced significant reductions in symptoms. It is important to note that it did not result in complete recovery as problematic symptoms were still present, but symptom reduction was sufficient so that 72.7% of NFB subjects no longer met criteria for PTSD.

Their focus on increasing alpha and inhibiting other frequencies is actually very traditional, and working at P4 and T4 was also a common practice for over a decade in NFB for trauma and anxiety. It was consistent with Gapen's opinion that alpha training was affecting limbic structures. They inhibited 4–7 and 22–36 Hz with minimal understanding of what the beta frequencies represented. They enhanced 12–15 Hz, which was a peculiar selection of frequencies, but perhaps based on the clinical observations of the Othmers and Sebern Fisher. In Gapen only, adjustments were made to the reward frequency in 1 Hz (down) to ½ Hz (up) increments based on symptoms of under or overarousal. This was an early Othmer approach prior to their shift to infra-slow training protocols and commonly used in the EEGer community. Training time in this study was gradually increased from 12 to 20 minutes.

The authors in this van der Kolk study noted Ros et al. (2013) had demonstrated that NFB increased functional connectivity within the alertness salience network involving the dorsal anterior and mid cingulate. This is similar to the frontal activation noted by Nicholson et al. (2020) and Ros et al. (2013), who had demonstrated that their NFB protocol had generated greater calmness, and greater salience network connectivity of the right insula, as well as enhanced DMN connectivity with the bilateral posterior cingulate in the posterior region. It also generated greater connectivity in the right middle frontal gyrus and left medial prefrontal cortex. The authors in the van der Kolk study also noted that in previous research, the frontal region was hypoactivated, reducing inhibitory control. They speculated there was a "cascade effect" throughout the network systems, which contradicts their localization theories. The idea that trauma was localized, has only one manifest pattern, or could be resolved utilizing only one localized location, is inconsistent with everything we presently know about the brain. They did, however, acknowledge that protocols may vary based on specific individual symptomology, and a one-size-fits-all approach would not make sense. They noted as well that clinical symptomology was an important aspect of protocol considerations:

"Adjustments were made based on rated symptoms of over-arousal (including nightmares, sleep difficulties, hyperactivity, aggressive behavior, anger, anxiety, and self-reports of high arousal including self-harm, suicidal and/or homicidal ideation), and symptoms of under-arousal (including inattention, decreased alertness or mental clarity, nausea, depressive symptoms, and decreased energy/fatigue)."

The authors comment that lack of self-regulation has been identified as a principal cause for the failure of exposure based treatments. They speculate that because of NFB's demonstrated ability in this study to generate statistically significant improvements in affect regulation, identity impairment, abandonment concerns, and tension reduction, it is a very strong candidate for treatment. This is especially true for those who are too anxious, dissociated, or dysregulated to endure other treatment modalities.

While all of this is fascinating for those who study such things, it must be kept in mind that we are only beginning to faintly understand the impact of trauma on human neurophysiology and the different ways it manifests. There will be decades of research ahead that will likely outline all the various ways that it impacts the brain and CNS at the individual level. What is of great value is to see the research demonstrating the ability of NFB to change the brain and CNS, especially in areas where other imaging techniques such as fMRI find unusual activity related to trauma. It is a critical validation at both the scientific and clinical level. As Walter Freeman said (Freeman et al., 2009), we need to consider these two technologies together moving forward. EEG neurofeedback, however, is far cheaper and likely just as effective as fMRI neurofeedback. They both show the power of feedback to alter structure and function of what appear to us to be key large-scale networks. Freeman also said that we should be looking at the brain from a systems perspective and using a form of field theory (Freeman, 2005).

This brings us to the front door of AT training. Not only do these studies validate NFB, but the research specifically points in the direction of AT training for those who have studied it, understood it, experienced it and deployed it clinically. We don't need this kind of sophistication with respect to network theory to do NFB, just to validate it and best mobilize it at the clinical level. We are training the mind-body system and we are not forcing it to change but inviting it to change. It reminds us that the human nervous system has a tremendous capacity to heal itself, but only if we can provide the right kind of context and support.

Protocol Considerations

The elevated activation of the DMN in fMRI terms translates to elevated posterior alpha frequencies in the EEG. Again, fMRI activation in different regions translates into different patterns of EEG depending on the dominant frequencies and related functions of those regions. Down regulating the PCC at Pz reduced activation in the bilateral dmPFC and hippocampus, postcentral gyrus, amygdala, cingulate cortex, temporal pole, middle, and superior temporal regions in salience areas related to emotional arousal, emotional evaluation, salience monitoring, and fight or flight mechanisms. The SN includes amygdala and BNST.

There is a strong relationship electrophysiologically between amplitude, phase, and ultimately coherence activity in the brain. Nicholson and his associates (Nicholson et al., 2020; Ros et al., 2013) describes their NFB procedure, alpha downtraining, as alpha desynchronization because it reduces the phase synchrony required to increase alpha amplitude (Nunez & Srinivasan, 2006). Consequently, reduced alpha amplitude decreases alpha synchrony and also connectivity within the DMN hub. This should in turn increase connectivity between anterior and posterior DMN as explained earlier. This activity should then theoretically normalize DMN and CEN activity. Nicholson et al. (2020) demonstrated that one session of this alpha inhibition at Pz, which desynchronizes alpha, has been shown to reduce DMN activity and activity in the amygdala as well as related physiological arousal levels.

Nicholson et al. (2020) also cites previous studies demonstrating that directly down-training the activity of the amygdala directly increased dlPFC and vlPFC activity and also resulted in many undesirable abreactions. This more indirect method of training at Pz which they employed resulted in significantly fewer abreactions in comparison.

This is of very significant importance for NFB training theory because it is well known that directly training the amygdala through temporal lobe locations of T3-T4 typically results in strong clinical abreactions. It helps to explain why P4-T4 down-training activity has calming effects with fewer abreactions.

The power of this finding is that it demonstrates that positive changes in the entire large scale network system can be achieved through different methods with either similar or different outcomes. It also suggests that alternative locations and methods may be even more robust, which they have not studied in such detail. One excellent candidate is AT training, which has a long established and well researched efficacy rate to support that supposition. AT training also results in posterior alpha desynchronization as well as both enhanced limbic theta inputs in the posterior theta pathways and enhanced frontal theta pathways. Key research has demonstrated that frontal theta is related to memory search as well as error detection and insights. These functions involve the same frontal and posterior Brodmann areas as the alpha posterior inhibits. Abreaction appears to be more common in AT training, but I would argue superior integration occurs as well.

It is very difficult to train down high amplitude posterior alpha during NFB. Bill Scott resorted to alpha inhibiting at Pz prior to AT training as a solution. This solution turned out to be effective. Clinically, Scott already anticipated this solution and Nicholson confirmed it. But it is only half the solution. It provides stability and some integration, but not at the level that AT offers with respect to complex trauma. Therefore alpha down-training may be a good solution for PTSD but AT may be superior for complex trauma.

Below is a pre-post magnitude headmap map showing the wide-ranging effects of AT training. The response to the protocol has been a shift from light blue and dark blue in the alpha and beta frequencies, representing one and two standard deviations below the norm, to green, representing the normal range of EEG activity. The brain has recovered its energy and ability to generate high frequency EEG and to operate with normal levels of alpha synchronization. This is systemic response to the protocol and did require targeting norms in various regions of interest regions of interest (ROI). It also did not require the alpha down-training as presented in Nicholson's (2022) experiments.

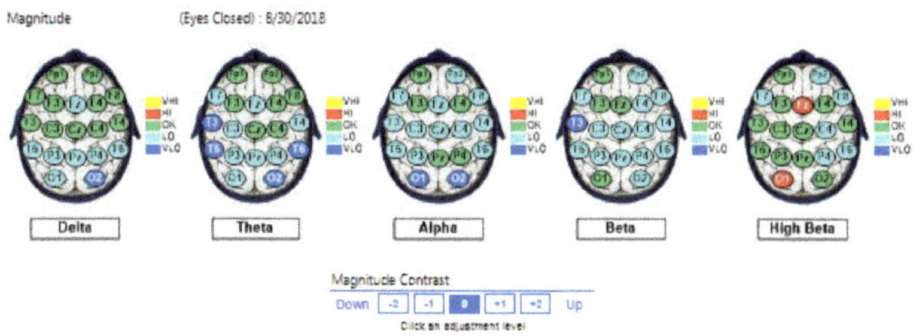
Magnitude
(Eyes Closed) : 8/30/2018
VHI
HI
OK
LO
VLO
Delta
Theta
Alpha
Beta
High Beta
Magnitude Contrast
Down -2 -1 0 +1 +2 Up
Click an adjustment level

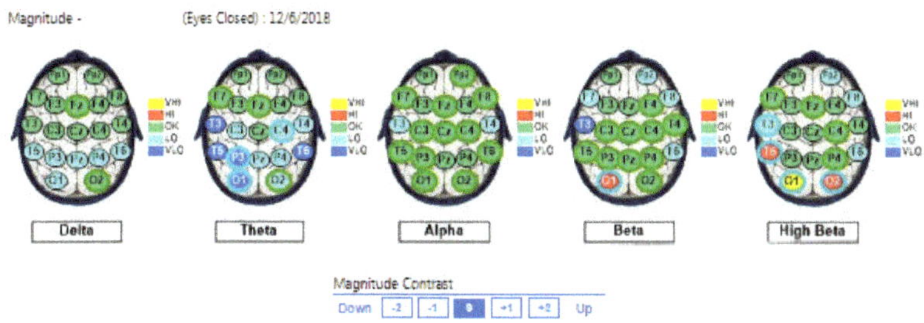
Magnitude -
(Eyes Closed) : 12/6/2018
VHI
HI
OK
LO
VLO
Delta
Theta
Alpha
Beta
High Beta
Magnitude Contrast
Down -2 -1 0 +1 +2 Up

Chapter 7

Using Neurofeedback With Trauma

Using NFB with trauma requires a two-phase process and has evolved along separate lines from formal complex trauma protocols but with many similarities. The initial phase, traditionally known as fast wave training (Crane & Soutar, 2000), is done to stabilize the individual's CNS and begin to calm the system down, lowering arousal and reducing chronic sympathetic activity. It is also applied to enhance sleep and sleep quality. This phase additionally allows the clinician and client to build rapport as the client learns to trust the NFB process and find value in the training. In this respect, the goals are similar to the ISTSS phase one model, and it evolved clinically for similar reasons. With enhanced CNS and emotional stability, experience in the efficacy of NFB, and improved resilience and plasticity that arises from the training, clients are better prepared to weather some of the difficult images and feelings that will emerge during AT training, as they become exposed to the original cognitive, emotional, and sensory components of their trauma-generating experience(s). This significantly reduces client dropout that sometimes occurs due to the intensity of the integration experience.

When dealing with clients with less severe trauma, many issues will come up and resolve themselves naturally as clients feel less depressed and anxious. Clients become more flexible and adaptive in their behavior and become less emotionally reactive to triggers. They experience improvements in sleep, cognitive functions, as well as emotional resilience and stability. These clients with less trauma can then be easily transitioned to AT training in 12–15 sessions. For those with more severe trauma, the clinician will observe more difficulty with the client's initial progress with constant backsliding in the first 15 sessions which often gets progressively worse toward the 20 sessions mark. The less severe clients are likely to achieve significant resolution of their issues in 30–40 sessions. The more severe cases can easily go into 60 or more sessions. As noted in many of the pre-post assessments done in the research previously discussed, major changes do occur, but the process rarely results in complete remission. As van der Kolk notes, even with the best methods such as CBT and exposure therapy, the majority may experience reductions in PTSD symptoms, but they rarely restore the client to pre-trauma levels of social engagement and enthusiasm for living (van der Kolk, 2014). Most have continued difficulty with their emotional states, their work, their relationships, and their health. In one of the largest studies done on CBT (Scnurr et al., 2007), most still suffered symptoms after 3 months, and only 15% no longer had major symptoms.

Another issue that arises during the initial training is the level of medication of the individual. More severe cases usually come with higher levels of medication as well as multiple psychoactive medications. Polypharmacy is frowned upon in the field of psychiatry and its associated journals, but it is widely practiced (Whitaker, 2010; Stahl, 2012). These drugs tend to block access to emotions, and we have noticed in hundreds of clinics over the last 20 years that they appear to significantly slow progress. Antipsychotic medications such as Risperdal, Abilify and Seroquel tend to dampen the emotional brain (van der Kolk, 2014; Thompson et al., 2019). This blunting effect also extends to SSRIs, and they can make enduring changes in the brain that make it difficult to function at a normal level of emotional self-awareness (Langley et al., 2023). This makes it very difficult to integrate past trauma.

Without this emotional component it is difficult for trauma to surface during sessions—and that is the intended purpose of the drugs, i.e., management. Yet at the same time, there is evidence that the intensity of emotional experience is a key component of the therapeutic integration process. Part of the process of stabilizing clients during the initial sessions is also to begin reducing their medications. This must be done very gradually, often more gradually than most psychiatrists tend to do. We have found in general that usually at least 3 weeks per reduction is essential, or clients tend to have severe abreactions leading to further compensatory medications. This is because of the rebound effects often caused by withdrawal.

In addition to these concerns are the cumulative health issues that must be confronted, properly managed, and stabilized. Training people who are in constant pain and suffering poor sleep, consequently, is very difficult. They tend to either remain at too high of an arousal level due to the pain and discomfort they are experiencing, or they tend to descend too far into sleep-like states. This can confound research outcomes as well and is usually not taken into account. Based on the findings of the ACE Study (Shonkoff & Garner, 2012) at least 60 percent of the clinical population will have significant health problems. This is one of the reasons the New Mind Assessment system includes a statistically validated physiological checklist that provides a clinical norm reference so that clinicians can determine if a client's physiological state is significantly deviant enough to undermine the efficacy of the NFB training and diminish the quality of the outcome. Clients with reference numbers two standard deviations above normal should be guided to a functional medicine specialist for treatment to help reduce physical symptoms contribution to their pain and discomfort, as well as any dietary issue that may be disrupting their sleep.

Typically, clients begin with a protocol derived from a qEEG brain map. At New Mind, we use mostly two-channel standard NFB. Our qEEG assessment system automatically provides a protocol based on the global EEG distribution and associated neurometrics that can significantly normalize deviant features and enhance neurological plasticity, stability, and resilience. The details of this approach are beyond the scope of this text, but I would refer interested readers to an entire chapter devoted to the theory and method behind it entitled "Perspective and Method for a qEEG Based Two-Channel Bi-Hemispheric Compensatory Model of Neurofeedback Training" (Soutar, 2017b). For most clients, this stabilization process takes about 15–20 sessions, at which point they are usually prepared to do AT. Bill Scott's CRI Help study as we mentioned earlier used a standard protocol twice a day for 10–20 days, but we find the two channel protocols, with enhanced nonvolitional proportional feedback methods we developed, accelerate the outcome for most clients. Once they have completed the initial sessions, they are prepared to engage in the AT training process.

Trauma Theory and NFB: The Process of Trauma and Integration: Memory Storage and Retrieval

Memories are not stored and recalled in a whole form and then played back as in modern video formats. Instead, each time we recall an event we reconstruct it in real time from components sorted in different sensory modalities. Each sensory system apparently contains its own memory stores, and when we recall an event each sensory system contributes its component of the memory (LeDoux, 2015; Mather, 2007).

Exactly how the brain sorts and binds these representations in real time is not well understood at this point, but awareness of the nature of this process has important implications for understanding trauma and PTSD. Information the brain encodes from any event has associated positive or negative emotional aspects, which is often termed emotional valence in research literature. It stands to reason that if an event has emotional valences which result in too high a level of arousal, such that it would manifest as panic, then the memory would never complete its binding process when recalled. This incomplete binding apparently generates a tension in the system that continually drives the system to resolve the memory, but the memory cannot be resolved.

Joseph LeDoux (1996) provides us with a model from his research on fear conditioning that supports the notion of a dual system of memory involving an implicit system and an explicit system. The implicit system appears to be subcortical, and generally employed in initially binding memories upon first encoding. The explicit system appears to be more cortically based and contains constructed narratives. He argues that implicit memories must eventually make their way to the explicit system to be fully integrated in the narrative form.

Narrative is a critical component for conducting and maintaining the ego and it is the primary routine task that engages the DMN of the brain. The brain appears to continually process memory information to maintain the ego, and the narrative is one of the key tools it appears to use in coordinating information. Some memories when processed result in too high of an arousal level verging on a state of panic. The amygdala becomes overwhelmed and in turn overdrives the sympathetic system. When this occurs, a memory cannot be resolved, and it constitutes an anomaly in the system. This is analogous to trying to digest food that has gone bad.

One example of this is the patient who had a dissociative disorder due to severe trauma and abuse in childhood. She initially reported strong smells occurring during alpha theta training. Several sessions later she reported hearing voices and then strange tactile feelings in different areas of her body as if someone were touching her. A few more sessions and a visual image emerged along with the other perceptions and converged into a terrifying memory of abuse. Then pieces of an entire scene and series of events developed, and she was able to recount the incident. From that point on we were able to assist her in integrating it into her other memories of childhood. This pattern became typical of what we observed fairly often in cases of childhood trauma resulting in PTSD. It also matches what LeDoux's research predicts.

Depression is like being in the belly of the whale, because the consequence of pharmaceuticals is to suppress emotions and therefore permanently close the mouth of the whale. Without the capacity to integrate, individuals are marooned in a state of blunted emotions and hopelessness. When given a couch, insomnia, snacks and a TV, the whole situation snowballs. Isolation becomes an emerging problem and negative self-talk becomes progressively more persuasive. The ensuing stress only enhances the inflammatory process. The individual becomes a helpless and hopeless passenger in his or her own drama.

The receptor levels in the synapses depopulate to a level that adjusts for the effect of the drug. Emotional development diminishes because access to emotions is permanently limited. As we train individuals suffering from depression at New Mind with neurofeedback, we find they become progressively irritable and restless, as they begin to move out of the depression into the original fear and anxiety that resulted in the depression. If the medication is not reduced, they become stuck. The response of most psychiatrists who do not understand this process is to prescribe more medication.

It confuses many clinicians to see clients shift from depression to anxiety, and it is difficult to understand based on their training. The confusion stems from a lack of information at many levels. First, depression was considered an aspect of anxiety until drug manufacturers developed Prozac and persuaded psychiatrists that depression was a completely separate category of disorder from anxiety. From a physiological perspective this perspective is wrong, but it helped boost drug sales significantly (Healy, 2004).

From a physiological and electrophysiological perspective anxiety and depression exist on a spectrum and overlap. As we have documented in the chapter on physiology, anxiety generates excessive excitotoxic activities in the brain. This results in a buildup of glutamate that is toxic to both astrocytes and neurons, but astrocytes can convert it to glutamine to be recycled in the neuron. At the same time norepinephrine levels increase in the right hemisphere. However, toxic levels overwhelm the astrocytes, and this triggers a proinflammatory response. It is this response that triggers the kynurenic pathway which converts 5-HT, the precursor to serotonin, into quinolinic acid to dampen the inflammatory response (Miura et al., 2008; Correia & Vale, 2022). Serotonin levels drop as an outcome to the anti-inflammatory response, at the same time sleep deprivation increases, which limits removal of glutamate and other toxins which further aggravates the situation. The reduction in serotonin in turn triggers further physiological changes in the endocrine system and the gut that shift individuals into depressive symptoms (Callaghan et al., 2020).

The electrophysiological response follows the neurochemical response. Initially elevated norepinephrine increases in the right hemisphere, resulting in increased beta frequency activity which becomes apparent in the qEEG (Engels et al., 2007). Inflammation begins to reduce ATP output which in turn reduces sodium and potassium pump activity and thereby diminishes the capacity of the brain to generate higher frequency brainwave activity. The beta frequencies become less prominent and alpha waves begin to appear more frequently as the brain is forced to idle due to lack of resources. Clients become foggy brained and alpha frequency activity begins to slow as serotonin levels drop even more. The alpha also begins to become more apparent in the left hemisphere region and diminished in the right hemisphere region resulting in an alpha asymmetry. Individuals begin to become more irritable and progressively withdrawn (Davidson, 2000; Solomon et al., 2014; Gold et al., 2012). When clients become excessively depressed an interesting compensation phenomenon can be observed. Beta begins to shift back to the left hemisphere. Clients with very high levels now have elevated alpha and beta in the left hemisphere. This has been statistically demonstrated to be a consistent pattern across clients (Soutar, 2021). Choi et al. (2011) as well as others have found that training to shift this asymmetry back to a more normal pattern reduces depressive features. Unfortunately, at the same time client irritability increases along with anxiety as they retrace the journey into depression with beta moving back to its dominant position in the right hemisphere.

At this point in the neurofeedback process clients begin to have vivid dreams and sometimes nightmares. This stage marks the beginning of integration of the emotionally intense or traumatic memory segment into the explicit memory system. This is the shift from implicit to explicit memory outlined by Joseph LeDoux (2015). Early studies in learning showed that rats would consolidate learning of mazes when a 7 Hz signal coupled the hippocampus and the frontal lobes during sleep (Buzsaki, 2006). If this part of their sleep was blocked, they could not learn the maze. Further studies suggested that theta played an important role during REM sleep in the integration of new learning comprised of associated memories (Rasch & Born, 2013; Cowdin et al., 2014). This demonstrates the importance of dreaming and REM sleep in the learning process. More recently, some have found evidence that both trauma and how it is handled by the brain are intimately related to REM activity (Murkar & De Kininck, 2018). Indeed, Sopp et al. (2019) found that longer REM periods after trauma were associated with better integration resulting in diminished symptom intensity. AT results in states analogous to REM with many similar characteristics, suggesting why it may be so effective at helping individuals integrate trauma.

What is rarely commented on is that most individuals with a mental disorder and on medications do not remember their dreams (Naiman, 2017). This may be due to the fact, which has been known for some time, that medications, especially SSRIs, reduce dream recall (Pace-Schott et al., 2001). This has been readily observable at the clinical level and hard to miss if you ask your clients about it. Based on the literature, it is especially difficult for medicated individuals to gain the proper amounts of REM sleep to integrate. Our clinical notes over 20 years consistently show that clients begin to dream more during NFB. As they reduce medications, they accelerate in their clinical recovery, whereas clients who refuse to reduce medications frequently respond poorly to the feedback process, as measured empirically by statistical analysis of their trend screens. In addition, they make a slower recovery qualitatively as measured by the symptom tracker. The pre-post maps confirm this outcome.

Given these findings it would be expected that individuals who are depressed are not likely to be able to integrate memories and their associated emotions if they cannot feel the emotion. Emotional valence is a key component of memory processing. Drugs like Prozac and other SSRIs that are known to mute emotion consequently block the oscillatory shifts between overarousal and the inhibited state of depression that modulate the gradual effort of an individual to process trauma and tragedy. This constitutes a reduction in both plasticity and compensatory degrees of freedom. This leaves clients stuck. Gradually reducing the SSRI during this phase allows the individual to dream and process more freely to integrate trauma.

The importance of sleep for mental health and recovery cannot be overstated and poor sleep is perhaps one of the continually most underestimated sources of disorder in the brain (Walker, 2017). Once the individual begins to normalize sleep, begins dreaming, and shows signs of lifting mood with episodes of irritability early in the training, it is time to shift the focus in fast wave training to aim at reducing arousal or alpha theta training to target trauma. The electrophysiological marker for this is the shift of alpha dominance to the right hemisphere and beta dominance to the left hemisphere. For those with moderate to light trauma on the developmental trauma spectrum, this typically occurs around sessions 10–15 but for more serious trauma this tends to occur in the 20 sessions or more range as clinicians begin to combat intermittent backsliding. This backsliding itself is an indicator more intense levels of emotional trauma.

With two-channel bi-hemispheric training, this shift in asymmetry is easy to observe with a good trend screen designed properly with this purpose in mind. The trend lines on the review screen appear to merge closely together and with some clients there is a clear shift. Unless your software is designed to detect this, it is difficult not only to observe but also to track. This is the importance of having software formats designed based on research findings tied to clinical experience rather than customer convenience and the typical engineering efforts to provide all possible outcomes. The trend lines converging and displaying the normative pattern of amplitudes for eyes open or eyes closed indicates increased plasticity, improved emotional flexibility and resilience, and increased communication between hemispheres as well as between cortical and limbic processing.

The client begins to emit novel behaviors in social contexts that previously would have elicited rigid stereotypical and automatic responses that lead to social error and inaccuracy and result in distress and conflict. These new behaviors result in innovative interactions with others leading to more positive outcomes. These behaviors reflect an integrated plastic neurology that can quickly adapt to changes as it responds creatively to challenges. But it is also usually a period of increased anxiety and agitation, and sometimes irritability, as the clients learn to trust their new responses. This period lasts a week or two but if it lasts longer, it indicates more pronounced chronic underlying anxiety based in unresolved trauma. At this point, alpha theta training is initiated to calm the CNS and uncover and integrate deeper and older traumas.

Alpha theta training is a much-misunderstood process in the field of neurofeedback, and many clinicians shy away from it. Others vastly oversimplify it. They see it as simply another form of operant conditioning that magically calms people. In truth, it is more of a method of using EEG to guide an individual into a specific state that we are capable of learning, but culturally discouraged from engaging in. Remember, this state, according to Elmer Green, was one that was part of yogic training and he had learned it from his yoga teacher. It was one of many meditative stages that individuals cycle through in the course of their yogic training, and he found it especially good for developing creative solutions to problems (Green & Green, 1977). As we learned in previous chapters, Green found that the key signature of this state when monitoring the brain with his EEG equipment was very high amplitude theta frequency in the area of O1 in the left occipital regions. This is perplexing from an anatomical perspective because O1 is part of the Brodmann areas 18 and 19 whose cell architecture develops around visual processing. In contrast, the God Spot observed by (Newberg, 2001) was at P4, a region which processes the boundaries between physical self and others and would seem a more likely candidate.

One caveat here is that it turns out that O1 and O2 at the back of the scalp have short and rapid pathways to the amygdalae (Ousdal et al., 2014), and become highly active when novel stimulation is present. This region stands out as highly deviant, when people are anxious, as a dysregulated region. This is one of the reasons our database generates protocols that target this region in highly anxious individuals. Our database operates based purely on statistical distributions, and at first, we thought it was an error, until the research confirmed it and by happenstance increased the validity of our methods.

In particular, theta it is statistically lower in the posterior region when individuals are highly anxious. Ironically, our database has been criticized for showing so many people with low posterior theta without consideration for the fact that most people with disorder have significant amounts of anxiety. It is therefore very interesting that Green, who had no access to such research at the time, recorded the opposite phenomenon of increasing posterior theta among many of those trained in yoga on his visit to India. Today we know from AT training with digital filters and sophisticated software that this is usually actually slowed alpha. His stated goal with alpha theta was to provide a more technologically measurable way of achieving this state with minimal training so the average person would have easy access.

As outlined in our AT history section, a small emerging community of practitioners adopted this approach to treating trauma. Many clinicians at Menninger, such as Dale Walters, continued to conduct workshops and contribute to the development of conferences exploring this new technology. Walters utilized a full peripheral biofeedback hookup as did Green. Many of us continued that tradition, although I would argue at this point from clinical experience with all of these technologies that HRV covers most of this dimension of training in one simple format and app. With the passage of time, we found the neurofeedback component to be a critical factor in the process, but the peripheral biofeedback measures were also important for monitoring progress. Adam Crane, who was my co-author on one of the first books on AT training, built his equipment and his workshops as well as his company around this psychophysiological perspective, and incidentally trained and sold equipment to Nancy White in Texas. Nancy eventually wrote the first published chapter on the theory of AT (White, 1999). HRV is progressively gaining widespread usage in NFB, but we have integrated it into our whole approach at New Mind and use it as one measure of integration.

AT training has proven effective, as I have mentioned previously, only with the addition of group meetings such as AA as Scott discovered in his research, but those with PTSD without addiction, especially with milder trauma, respond well in the usual clinical setting. It is so effective that it has a profound impact on trauma integration, and this has been acknowledged specifically in the research findings that put it in the probably efficacious category (Scott, 2017). In general, AT is very effective in reducing the startle response, which was until recently thought impossible to change. AT is highly effective in reducing general sympathetic tone, which can easily be empirically confirmed by monitoring with biofeedback equipment with almost any given client. Since many of us were using full peripheral biofeedback hookups along with NFB early on, it was hard to ignore. It led many to emphasize NFB over peripherals, and lead to the division between AAPB and ISNR which still exists to this day. This reduction in sympathetic tone contributes to reductions in insomnia and anxiety. It also has an impressive impact on a suppressed immune system and the hypothalamic pituitary axis (HPA). Normalization of menstrual cycles and hormonal issues are not uncommon, and it can have a strong impact on PMS. With the increased immune function, many people experience increased energy and relief from a variety of symptoms with unidentified sources.

The ability of alpha theta training to guide clients into a unique twilight state allows them to remain at a very reduced arousal level that appears to facilitate the integration of memories stuck in the implicit memory system, that would typically result in such high levels of overarousal and panic that they cannot be consolidated and integrated. This was the main thrust of Nancy White's theoretical perspective outlined earlier. In the long term, the integration of these traumas results in increased integration of neural networks and facilitation of neuronal function. The partitioning of these networks due to trauma appears to diminish the functionality of the system by blocking access to processing resources in limbic structures, due to gating in efferent and afferent pathways between the cortex and the limbic system. The suppression appears to be along the lines described by LeDoux (1996; 2015) in which worry functions as a gating mechanism between the orbital frontal regions, key executive processing areas, and the amygdala. Individuals then have minimal access to their full emotional resources and are not as efficient at emotional valencing. Decoding social contexts becomes more challenging and their rate of behavioral error increases, reducing their social and interactional accuracy, generating even more distress. This escalation of distress then elicits more emotion and arousal, which in turn places pressure on the system to respond at a higher arousal level in an escalating cycle. The limbic response then begins to assemble the traumatic memory again, and again panic ensues. This vicious cycle continues indefinitely until the trauma can be resolved. In addition, any event that is like the original trauma can also induce the same process and consequently minor events can be a trigger for this process. In a sense, integration is a drive derived from the dynamics of the entire neural system in its efforts to manage its social ecology and maintain internal allostasis. Social cues can carry existential threats.

For the average individual, lower levels of trauma appear to have a less intense cycle. More manageable but still significant traumas occurring during the socialization process trigger stereotypical and scripted responses that induce stereotypical counter responses based on violations of social norms and mores regarding appropriate conduct. A cascade of automated responses and counter responses results in escalating emotional tension which ends with conflict or escape from the context, i.e. fight or flight. This automated response pervades our entire biology and is a key component of biological organization. Without routine and habituation, biology cannot exist. This concept is covered in more details in my book The Automatic Self (Soutar, 2015) and is covered in even more details in my associated lectures on LifeStatesTechnologies Youtube channel via http://tinyurl.com/2w8b35um.

What binds individuals to these scripted responses is the constant conditioning that shaped them in the family of their origin. At another level, they are based on expectations and erroneous untested predictions about the world and how it functions. The predictions reinforce their view model of their "Lifeworld" and in a sense, provide a false comfort that the world is as expected and predictable. This sense of security is what maintains their trances or ongoing self-talk which in turn generates mindsets that continually define their Lifeworld. Mindsets arise and aggregate around fundamental beliefs, values, and biases derived from their family of origin that individuals are only vaguely aware that they carry. Examples might be all men just want sex, or people are not to be trusted, it's a dog-eat-dog world out there. These can extend into the domains of politics and religion as well, and now they are being amplified by anonymity in social media.

Mindsets are often composed of several trances that take turns in repetition cycles. These usually start out as mind chatter and help block and manage past emotional trauma (Ros et al., 2013), and typically present as elevated posterior alpha above 18 μV (except in cases of globally low power EEG). Many of these can start out as worry in the form of overactive frontal activity showing up as excessive beta waves and thought categories of "what if?" If they become too obsessive, they may shift into higher amplitude frontal spindling beta. In the figure below is an example of beta spindles in map and below that the raw EEG.

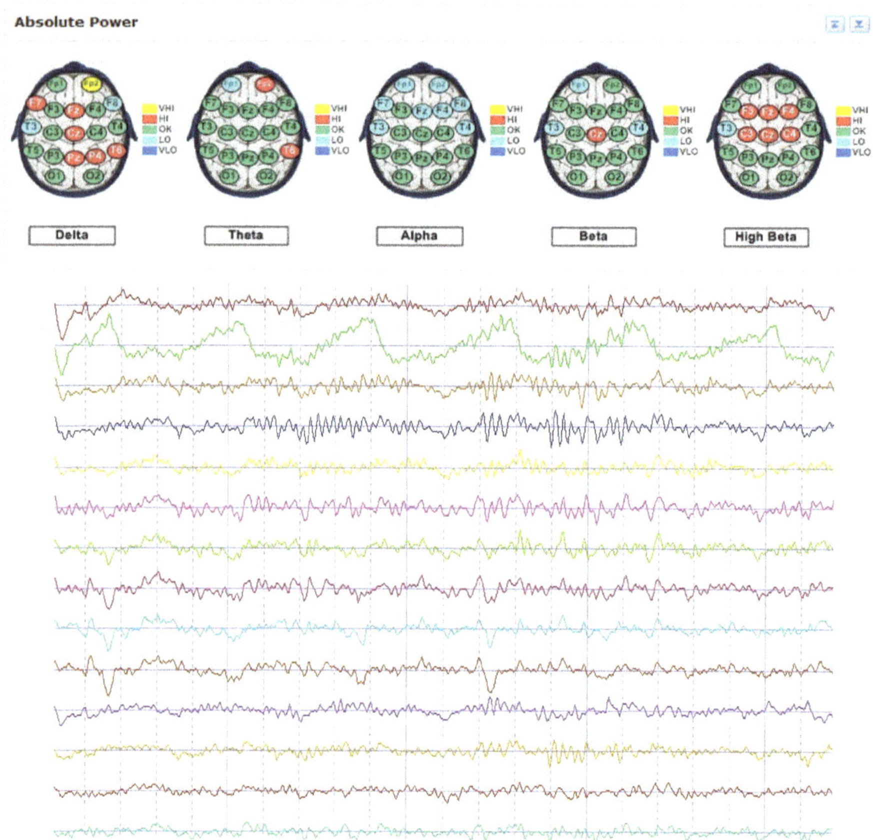

These trances are also sustained by their routine maintenance involving repeated imaginary scenes which can escalate to constant negative ruminations which often present as posterior elevated beta (Engels et al., 2007). These routine and repetitive thoughts fall into clearly distinguishable patterns such as catastrophizing, minimization, black and white thinking, tunnel vision etc. They are an outgrowth of the fast-thinking component of the brain that progressively ossifies into a rigid repetitive pattern reflecting highly established patterns of neural metastable networks.

Another method of maintenance of emotional management is repeated behaviors which can involve subversion of self-awareness into tasks such as work manifesting as workaholism, or into the extreme repetitive tasks such as checking in OCD or trichotillomania. These later behaviors represent escalating forms of worry and rumination that trigger ritual repetitive motor activity. The lesser versions of these are nail biting, facial tics, excessive blinking, etc., known as adapters. Thinking is considered an extension of motor activity as it evolved out of gesture, especially facial expressions, and language (Prieur et al., 2020). One only needs to inspect the brain homunculus to observe how much dedicated processing space is devoted to the face and hands. This is likely an evolutionary adaptation related to intense social co-operation. Emotion was an important component of gestures. Conversion of emotional intensity that needs to be managed can take the form of physical pains related to skeletal and smooth muscle bracing, excess stomach acid and gas pains as well as GI spasms, etc. This phenomenon is, of course, Freud's observation of conversion, and it is well established but I do not believe it has been developed enough.

There are several dimensions to the developmental process that play key roles in shaping the individual and their personal experience and response to trauma. These dimensions also play a key role in the formation and development of the individual's brain. The structural adaptation to trauma requires extensive training and considerable time to change. It develops through the introduction of new inhibitory networks and the integration of subcortical abandoned memory parcels into a cohesive and manageable conscious verbal cortical narrative. It requires intervention on both levels, and LeDoux (2015) cautions that interventions that act first at the implicit subcortical level, like AT, may have the best results.

Clinicians and life coaches need to also be aware of these dimensions in order to effectively interpret the client's history and identify the multiple vectors leading to their symptoms and maladaptive behaviors. This awareness of the various dimensions enables the practitioner to effectively manage the client's responses to training and aids in supporting and guiding the client into uncovery, self-discovery, and recovery through insight and creative adaptation instead of fear based automated responses. The components of the New Mind perspective in what follows is but one way to characterize the components of human behavior, but it has proven very effective clinically over the past 30 years with pre-post psychometric and qEEG neurometric measures. It synthesizes decades of findings from many disciplines and correlates with the emerging qEEG findings and related clinical experience derived from three decades of AT training individuals from all walks of life.

Emotional Scripts

As Alan Shore notes (Shore, 1994), most of the families' emotional scripts are absorbed before the age of one and a half. Children begin imitating those patterns within the family context and by age 2 begin becoming aware of themselves as separate individuals and developing their own narratives. This is the point at which childhood memories become more enduring and simple self-narratives begin to develop, such "I can do it by myself." Children's physiological patterns reflect their emotional shifts that define their developing somatic landscape as defined by Damasio (1994). As children in highly abusive families develop in ways that facilitate and sustain them, their brain sizes are even smaller (Teicher et al., 2016). The EEG pattern of parents as well as their behavior is highly heritable (Peterson & Posner, 2012), and approximately 70% gets passed on but the strongest and most enduring patterns are in the posterior region (Zietsch et al., 2007). The frontal patterns seem more susceptible to ongoing socialization. This pattern correlates with the 50 percent nature versus 50 percent nurture model used in psychology to describe the contribution of genetics versus socialization to human psychological behavior. This would reflect physiological patterns that support the transference of behavior through vicarious learning and operant conditioning that occurs within the family system during the socialization process. As individuals age and engage in progressively larger and diverse social groups like school, camp, and cliques, their frontal lobes progressively develop and myelinate, with the prefrontal regions fully maturing at the approximate age of 25 (Niedermeyer & Lopes da Silva, 2005). The early emotional and simple verbal scripts begin to shape interaction with expanding group participation evolving in unique variants of the family of origin scripts. These variants may develop either positive or destructive social trajectories. Emotional scripts also are part of the fast-thinking process described by Kahneman (2011), which is highly tied to subcortical processing (and theta pathways described in two thetas). It is the emotional scripts that, when dysfunctionally developed, coalesce into the negative personalized patterns that Judith Beck outlines in CBT theory.

Core Dramas

During childhood, we see the emergence of individual core dramas that are forged out of the conditioning of the family context during the expression of the repertoire of family transgenerational emotional and behavioral scripts. Over time, unique patterns of response to different family scenes develop during development as each child carves out a role within the family system. The successful ones leading to attention or other rewards become well developed and consolidated into a group of automatic scripts that get enacted in response to each scene. The more traumatized the family, the more stereotypical, rigid, and repetitive the scenes enacted. In a sense, the family becomes trapped in its own repeating patterns as the family system tries different iterations of enactment to resolve the trauma of each individual. This creates multiple cycling iterations within the family constellation. The core drama becomes the central solution to resolution that never actually resolves. This same drama is triggered by situations outside the family and the individual is attracted to these triggers to find a unique solution. The only way out is the emission of a novel response. In addition, they are attracted because the triggers are familiar patterns that elicit behaviors that are well rehearsed, and the repeated emission of their responses reinforces their sense of identity. Most individuals prefer to have a defective identity rather than no identity, which can lead to anomie and suicide due to progressive social isolation (which one of the founders of Sociology, Emile Durkheim, demonstrated in the first social statistical study ever conducted and published).

Core Roles

As individuals become socialized into the family system, they tend to take on various roles in response to other dominant roles, and to accommodate those dominant roles. In time, they gravitate toward one dominant role more than the others and consolidate their identity, including scripts and stereotypical behaviors associated with those scripts (Bradshaw, 1988). Researchers have identified the most typical roles that emerge, especially in dysfunctional family systems. These include the Hero, the Scapegoat, The Rescuer, The Mascot, The Enabler, Little Mom, The Lost Child, and The Saint (Samuel et al., 2014; Wegscheider, 1989). The older child is often the Hero and always does the best and what is right. The Hero always follows the rules and is the center of the family's positive attention. The Lost Child is typically the middle child who is always in the background of family activity and receives the least amount of attention. The Scapegoat is always the center of negative attention and often competes with the Hero. Sometimes these roles can be equally split by twins. The scapegoat always distracts members from family activities with disruptive and oppositional behavior in order to garner attention. The characteristics of these roles are too complex to fully review here but there is extensive literature on this topic in family systems theory.

Scripts and Novel Action

The repetition of verbal scripts and related behavior confirms role identity and hence group identity. The response to verbal scripts can also be stereotypical and identity forming, such as when the parent says, "When will you ever learn, How many times have I told you, When will you get it through that thick head of yours, You're pathetic, How are you ever going to make it in this dog eat dog world if you don't pay attention to what I am telling you?" The list goes on. There is a conditioned emotional response in the child to these statements, and beliefs about efficacy and identity are conditioned as well. Scripts are automatic and effortless, and this generates a false feeling of authenticity. This false authenticity arises because the individual has not experienced an adequate number of events where they emitted spontaneous novel behavior based on their own feelings and from their own creative source that resolved tension, friction, or conflict. Their responses are always highly conditioned, automatic, and stereotypical. They become dependent on automatic family repertoires that result in stereotypical responses that are familiar and expected rather than relying on novel action based in self-awareness of feelings. Their hypervigilance results in global somatic bracing, showing up as elevated high frequency beta, that leaks into lower beta digital filters as well, and the EMG bracing that reduces emotional self-awareness and access to their somatic landscape (Damasio, 1999). This lack of access to a full emotional repertoire also continues to make them reliant on automatic scripts. The response is often seen in the EEG as significant theta reductions in the DMN and temporal lobes observed in the New Mind qEEG brain maps along with temporal lobe hypocoherence, which indicates a lack of connectivity in this key hub for emotional valancing. The lack of encounters with evoked novel behaviors resulting in successful interpersonal negotiations results in a lack of experience with the sensation of self-efficacy and self-empowerment that successful encounters resolve in. This results in a growing sense of frustration and disempowerment. If the familiar outcome of an automatic behavior is traumatic, then they may develop panic and rage to protect themselves.

Trances

Trances are like themes that people base their behavior on (Wolinsky, 1991). They are automatic behaviors stemming from fundamental beliefs about themselves. They may see themselves as a "loser" or believe they are a "good dresser." From these themes, they acquire a group of behaviors to compensate for, or enhance, these themes. These conditioned automatic behaviors are emitted in cycles that involve them with other individuals and institutions that can either be adaptive or maladaptive.

From a psychological perspective, they are partially dissociated, and from electrophysiological perspective they tend to produce excess theta waves globally as they are highly engaged in internal self-referential visual dramas, often in the form of rehearsal for future action. This is because their awareness is grounded in limbic functions more than cortical functions. They are in a daydream state similar to childhood consciousness before the brain increases its dominant frequency and begins to operate more cortically. This state is highly auto suggestive and has a hypnotic quality to it. Alcohol and THC can enhance this state. People shift into this state typically in response to stress, in order to generate an input barrier, and it can lead to intense episodes of daydreaming and eventually dissociation. They are a common childhood response to excessive family tension and conflict. The adaptors they engage in once they access this state are often self-soothing, and may include many compulsive patterns such as gambling, shopping, eating, sexual ritual, etc. They can also involve violent activities as well.

Trances vary in intensity and frequency, and they may be aggravated by TBI and emotional trauma. People often look back at their behavior and comment "I don't know what I was thinking," because they were engaging in behavior that is not rationally based. On the positive side, the ability to enter these states allows us to engage in play and pretend as well as co-operate in drama and the arts. This is why theta uptraining in AT work can often significantly enhance creativity in professional artists as well as creative decision-making in professional managers.

On the negative side, they can lead us into self-destructive behavior that is irrationally based and difficult to overcome. NFB can be especially effective at mitigating these negative psychological states and decreasing the automatic component that can become self-limiting and self-destructive.

Rationalizations and Boundaries

As Bradshaw (1988) notes, family and social contexts where continual violations of boundaries and failure occur generate high levels of shame and guilt in family members. Boundaries are initially tentative and highly permeable in very young children, but become more established and concrete with middle childhood. Adolescence then results in rapid cycles of individuation as the teenager attempts to establish their own identity independent of the parent, and thereby begins to develop an advanced sense of self-efficacy. The first steps of the child's entrance into boundary establishment is the word "No," which comes with the first sense of separation and self-awareness, as is well known in developmental psychology. By age, of 4 this "No" evolves into "I can do it myself," when tying shoelaces or other fascinating and challenging activities of daily living are mastered. I can dress myself, I can wash myself, etc. become milestones of accomplishment that children take pride in learning and teaching their siblings. Easily frustrated and traumatized parents can rush or invasively take over such tasks generating a sense of shame and frustration in the child.

Parents may challenge the child's decisions either supportively or destructively. "I see you have two different color socks on?" vs. "Oh my gosh can't you do anything right, we have to leave right away and look at you!" Out of the challenge comes a reason or rationalization for the behavior and often it is learned vicariously from watching adults or siblings respond to perceived similar circumstances. Responses can be silly or profound, and kids do say the darndest things that are often delightful, but the inappropriate response of laughing at the child instead of with the child can make the difference between confidence and shame. In any event, rationalizations evolve and become progressively sophisticated as children explore lying and passive aggressive solutions to confrontation and questioning. In teenagers, it is not uncommon for it to include intense sarcasm, provocative statements, resentment, anger, rage, and a host of other strategies to maintain control, power, and draw lines in the sand or boundaries.

The manifestation of high levels of frustration and suppressed emotion frequently results in disengaged family systems. The rigidly disengaged family constellation is common in addiction-burdened and traumatized families (Soutar, 1993; Olson et al., 1989). Chaotically enmeshed families, involving constant extremes of emotional behaviors such as borderline personality disorders and bipolar disorder, as well as constantly shifting and dissolving boundaries, usually don't last very long. In rigidly disengaged families, rationalizations and defensive responses tend to dominate exchanges and negotiations tend to result in a failure to compromise and find solutions. Automatic thoughts tend to dominate thought processes and distort the contexts of social exchange. This, in turn, biases perceptions and decoding of body language and facial expression. Feelings are hidden to protect the individual from negative responses, which contributes to practiced self-unawareness. Emotional boundaries are less defined, protected, and honored, but physical boundaries are often rigidly defined. Family members may avoid exchanges and isolate themselves in separate activities. Children fail to witness encounters that model the proper use of anger and frustration as well as the repair of frustrating encounters by adults. Vicarious learning of these solutions does not occur. Instead, they witness strategic methods to hide feelings, since feelings only provide an opportunity for manipulation, and they avoid encounters that lead to goal fulfillment in favor of goals leading to short term gratification.

Since conflict resolution takes time, negotiation, and multiple encounters with exchange of authentic emotion, are not modeled. Rather, angry responses that halt negotiation and reward the individual with the greatest power with immediate gratification become the identified preferred method of negotiation. This individual is usually the parent, but not always. These angry or avoidant behaviors are often repeated in school in the form of bullying. Distressed and isolated kids seek affiliation with others with similar norms as well as the formation of identity groups of cliques that engage in deviant behavior.

Rationalizations allow us to repeat ineffective behaviors because they attempt to recharacterize and minimize the negative impact they actually have on relationships. The rationale becomes more important than the relationship and evolves into a trance. Others become alter-casted, and cast into roles they don't endorse, and into identities that do not actually fit, but justify the response to their alter-casted role. "Jimmy is stupid and so he should be treated accordingly. Someone has to wake him up so he can see how stupid his behavior is, and I certainly shouldn't have to put up with it." In this circumstance Jimmy is alter-casted into the role of the village idiot and everyone knows he deserves what he gets. This is the kind of rationalization that generates the Scapegoat role in children subject to this type of rationalization coming from other family members.

Sociological Contexts

Social forces are external and coercive. They exert pressure on social players in any context in a fashion that has both psychological and physiological effects, through the mechanisms of social distress (Alkadhi, 2013; Yuen, 2012). This includes the often-overlooked dimension of occupational stress (Savic, 2013). These effects are often minimized by those who are unfamiliar with sociological literature despite over a century of detailed analysis, experimentation, and documentation. Events such as divorce or the addition of a new family member to a home are among the most stressful in this category (Holmes & Rahe, 1967). These events have a very real effect on physiology and can result in stress leading to neuronal death (Sapolsky, 1999), unhealthy patterns of fat distribution and arterial deposits, or telomere shortening, and the consequent reduction in life span. Parenting style has even recently been demonstrated to have a significant effect on RNA and DNA (Peterson & Posner, 2012).

It has been more than convenient for policy makers in our social order to ignore and minimize these facts and exert pressure on others to ignore them. I can think of no other better example than how the NFL dealt with TBI for many decades. This pattern of ignorance has been epidemic in psychology, medicine, and psychiatry until recent times. The focus has instead been upon interpreting social distress and the resulting psychological issues as purely chemical cascades gone awry that can easily be managed with the appropriate magic bullet in the form of modern pharmacology (Whitaker, 2010; van der Kolk, 2014).

The oldest social group in our species recognized by anthropology is the extended family or clan. We have come to recognize the family as a complex system of ongoing interactions among individuals that has a dynamic of its own. The family shapes the individual as much as the individual shapes the family. Among the most powerful forces at work in the family is the socialization process that, over time, defines an individual's habitual responses to a given social environment. Forces such as operant conditioning, classical conditioning, and vicarious learning have been documented as mechanisms of action in this process (Bandura, 1997). Up to the present, the contributors to this shaping of the individual's collective repertoire of responses have been considered half nature, or genetic, and half nurture, or learned. The famous concordance studies of the twentieth century are among the many investigations that continually affirm this estimate. However, it has been reported that perhaps 70% of the EEG pattern that individuals display is heritable (Zietsch et al., 2007). It is interesting to note that the posterior region, most active in childhood, is most heritable, while the anterior region, most active in young adults into adulthood, is less heritable and more acquired. It is also well established that the pre-frontal lobes, most dominant in socio-emotional processing, are the last regions of the brain to myelinate. It is now recognized that most individuals have established the familial patterns of emotional scripts before the age of one and a half (Shore, 1994) and that most individuals do not begin a self-narrative pattern until the age of two and a half, usually commencing with mastery of the word "no." This emergence of such a self-reflecting individual is in a sea of mores, norms, and emotional currents that are profoundly complex. To cope with the complexity is apparently why we have evolved such large frontal lobes.

Children developing in families under the burden of profound social distress and/or poverty, the most correlated variables according to a century of demographic analysis, have deficits in both structure and function of their brain. They have smaller brains that are slower to myelinate (Kang et al., 2012; Jeong et al., 2021). This reduces processing speed, for starters. They may have oversensitive amygdalae and an exaggerated startle response from trauma. They are often provided with a narrow diet that is impoverished from a nutritional perspective. Diets tend to contain more contaminants and pesticides. They may grow up in geographical locations that are more exposed to environmental toxins. They are more likely to be exposed to drug addiction, alcoholism, and physical trauma. They are more likely to come from broken families and rigidly disengaged families (Soutar, 1993).

In terms of behavior, children from dysfunctional families are likely to learn emotional scripts involving transgenerational behavior that is maladaptive. They are not tolerant of delayed gratification and have low frustration levels. They are likely to anger more quickly, more likely to escalate to rage, and slower to self-calm. They are not practiced or aware of how to repair relations and it has not been modeled well in most instances.

The inability to delay gratification has been the focus of very interesting research regarding social deviance (Gottfredson & Hirschi, 1990). Individuals who are emotionally traumatized by dysregulated family systems find it very difficult to delay gratification. Providing a family with steady emotional environment that is predictable, based on clearly stated family rules, helps to emotionally support children through moments of frustrating challenges. It ensures them that their efforts will be accepted with patience, but within well-defined limits. Knowing that there are consistent limits provides a sense of security, especially if there are well-modeled guides to alternatives to action that can result in more successful outcomes. All these aspects of a healthy family contribute to the ability to patiently delay gratification, with a believable expectation of eventual reward and closure related to any effort. Knowing that failure will not result in a case of "sudden death" due to an adult overreacting, also helps children to understand that short term failure does not lead to long term failure, but to adaptive responses resulting in long term success. The child or adolescent learns to shift their focus from falling down, to the main concern of picking yourself up and adjusting your behavior accordingly.

Outside the family, school and work have a profound impact on the individual. We have repeatedly seen cases in our clinics of children emotionally traumatized by teachers, resulting in a global failure of academic performance. We have documented its impact on the EEG. In one case the child came from a healthy family system and received excellent nutrition but suddenly developed attention problems and began failing in school. The child showed high levels of global theta, particularly in the posterior region where the DMN resides. Efforts to down-train this 25–30 microvolts of theta proved futile which suggested, as is usually the case, that there was an unidentified confound present. Through consistent careful probing I was able to gain the young man's trust and discover that he had a teacher who harassed and derogated him daily.

I asked one of my fellow clinicians who still had contacts in the school system if she would contact the school counselor and investigate the situation. It turned out that the school was aware that this teacher was burned out and was frequently hard on many of the children. She was apparently close to retirement, and they were reluctant to let her go so close to retirement, as that would undermine her ability to receive her retirement benefits. So, we were able to arrange to have the child moved to another class and within a week the theta began to drop in power.

The stubborn high amplitude theta was the telltale indicator and we had seen it before. It is this kind of access to the ongoing physiological record that provides biofeedback practitioners data to see the immediate impact on the physiology and psychological status that other technologies or methods do not offer, and gains important insights to the individual's circumstances. The impact of this social situation had an immediate impact on the child's psyche and EEG, such that the child was withdrawing more into a fantasy realm to wall off the teacher's constant criticism and protect himself from the emotional trauma that was occurring. This shift inward was reflected in the shift to excessive processing activity in the self-referential DMN that occurs with dissociative fantasy. It is one of the many EEG patterns or subtypes representing the ways that individuals deal with emotional trauma.

I have also seen this pattern in adults very frequently, and in one particular case of a client who had chronic sexual fantasies he could not control. I was able to predict during his training, to his surprise, when he was training, exactly when he was engaging in the fantasies, and how much, by observing his theta activity. The frequency and distribution of the EEG amplitudes and frequencies are similar in these trance-like states to AT crossovers, but they differ in subtle ways. This appears to demonstrate the natural inclination in many people to dissociate in this manner to cope with social and psychological distress. My ability to monitor this client's mental activity in this manner inspired more confidence in him that we could identify the issue, and train him to reduce his dissociative activity. The next step, however, would be to uncover the source of his trauma through AT training.

Back to the child with the burned out teacher. The observed reductions of the abnormally chronic and high amplitude theta in the child over the next 2 weeks, and the dramatic improvement in his training, as observed on the trend screens, documented his response to the change. In the absence of any other discernable significant shifts in his activities of daily living, it became obvious that we had isolated the source of his presenting problems with paying attention in school. He did not require medication and he did not have ADD at all, but instead had been emotionally traumatized and his resulting anxiety and depression had derailed his motivation to attend school. It was fortuitous that we were able to catch this so early in the developing onset of his problems, and could identify the abnormality in his EEG pattern. Fortunately, his parents did not accept the diagnosis of a sudden onset of ADD, which was suspicious to begin with, and were quick to research other solutions beside drugs. As his brainwave distribution normalized, his symptom tracker documented his dramatic changes in negative behaviors and the positive emotional changes from neurofeedback, resulting in recovery of academic performance. His CPT testing also confirmed his improvements in attention and memory. In addition, we could see the change in his brainwave asymmetry in his pre-post qEEG, as alpha and beta shifted to dominance in their respective hemispheres, indicating his anxiety and depression had lifted.

The impact of abuse of authority in the workplace is also very well documented and its impact on the EEG is starting to be recognized (Sapolsky, 1999). Depression is a common outcome. The impact on health is profound. Yet, again, it is largely ignored at the public and policy level and especially minimized by corporate organizations. Nevertheless, it is a very real problem at the clinical level. We have many documented cases of severe depression arising from abusive bosses in circumstances where individuals feel they do not have economic alternatives with respect to employment. These cases have abnormal qEEGs that respond to neurofeedback training as well.

One case, in particular, involved a single middle-aged mother who had been working for years under a highly abusive boss. Not only was he verbally abusive but extremely passive aggressive and freely took it out on all of the employees. Working at this job and the complex trauma it generated, over time, had taken a progressively devastating toll on her and she had become profoundly depressed. The qEEG brainmap clearly showed a strong alpha asymmetry with excessive left hemisphere alpha. She also scored very high on both the Beck Depression Inventory and the Interactive Self Inventory I gave her. These two measures usually correlate quite well when asymmetry is measured using the New Mind algorithm.

She was reluctant to leave her job because of her age and education level and her concern that she would not be able to provide support for her daughter who had just begun attending a local college and was living at home. After several sessions we determined that her major concern was losing her home, which was her primary source of security in her life. Whenever the topic came up, her biofeedback measures showed dramatic shifts indicating extreme distress. As is typical in these cases, she was not sleeping well and had periods of extreme anxiety. Once we improved her sleep and stabilized her emotionally, we were able to commence reductions in her SSRI and her Wellbutrin, working closely with her psychiatrist who seemed surprisingly open to the idea. This would allow her greater access to her feelings and allow her to remember her dreams again. As we noted from the research review, the ability to access these to resources of the psyche are crucial to good progress with AT training.

We began training, and she complained that her anxiety was getting worse, and she felt more moody. This is typical of the first sessions and indicates good initial progress if the trend screen shows alpha diminishing in the first 10–12 minutes, and some small increases in theta. I reassured her that it was temporary and partly due to her backing off her medications as well as an indicator of good progress. After a few sessions she began to get crossovers and reported dreaming again with some of the dreams being very vivid. Within a few more sessions she had a very vivid dream that disturbed her quite a bit, but she was unsure why. She was in her kitchen, and she could hear puppies crying out from behind the basement door. She was worried about what would happen if she let them out and was unsure, she could manage all of them. She wanted to know what the dream was about, and I declined to answer asking her to think about it as she was settling into the outset of her session that day, and then let it go for the rest of the session. She was disappointed and frustrated at the end of the session and complained she still did not know what it was about. I told her to let it go and give it time.

I did not hear from her for a week, which was unusual because she was showing up regularly for her sessions. Then she showed up for her appointed session the next week and looked concerned and confused. She told me she was driving to work in rush hour through the pouring rain when suddenly she realized what the dream was about, and had to pull onto the shoulder because she was suddenly overwhelmed with emotion. She reported that she cried in intense overwhelming waves for 20 minutes and then was suddenly very calm and peaceful. She went home that day and called in sick for work. She asked me if she was going crazy. This is not an unusual question to encounter from clients when doing AT training. The intense emotional breakthroughs and experiences that follow are often novel, intense, and in some cases transpersonal. Her sudden and unexpected experience outside the office was not that unusual, either. Some clients do not feel safe enough in the clinical setting to experience breakthroughs and insights. They typically experience them at home or in dreams instead of having hypnogogic images and insights during the session.

When I asked her what she had decided the dream was about she acknowledged that the puppies were all her pent-up emotions, and that she knew she could not tolerate her work environment much longer. She was terrified that she would not be able to find work anywhere else, and that she would lose her house and end up financially abandoning her daughter who needed support to get through school. She was clearly stuck in what she saw as a no-win situation. Her real dream, she admitted, was to own her own gelato store, but she could not take the risk of leaving her job to learn the business, and did not have the savings to start up the store.

Sessions went on for a few weeks more with little progress and then I did not hear from her for about two weeks. She finally called the office and said she would be in at her regular time and when she showed up, she had a remarkable story. A tornado had swept through her neighborhood knocked down a tree by her house. It fell through the roof into her bedroom stopping a few feet from her face. Her kitchen was destroyed as well. She had been forced to move in with her neighbors who kindly extended an invitation while she fought to collect the insurance to repair her house. I commented that the decision to move out of her house appeared to have been made for her. She laughed in agreement.

This client continued to come in for sessions but admitted the process was starting to unnerve her when I explained the concept of synchronicity, as espoused by Carl Jung, and how it often emerged during training session with AT. Unusual coincidences often occur around key events related to the trauma. In fact, she collected enough insurance money to redo the interior of her house and tuck away a tidy amount in her savings account. Still, she did not feel safe enough to leave her job and try something else like the gelato store, but after a few more sessions she had another very vivid and disturbing dream she could not entirely unravel. In the dream she was back in her kitchen and opened the door to the basement and collected all the little puppies and put them in her car with her suitcase. They were driving away from the house to a new and wonderful place she had found, but suddenly a huge black bug flew in through a partially open window and hit her in the head. Her dream ended with her waking with her heart pounding in the middle of the night.

Once again, we did several sessions with little progress on her future plans and the same complaints about her boss. Her depression was lifting, and she went off her medications, but she was still very anxious and felt frustrated and stuck. In this wash-rinse-repeat cycle, I did not hear from her again for a several weeks. The next time she came to the office, she had another unusual story about suddenly having terrible headaches, and then going to the doctor and getting x-rays only to find a tumor in her brain. I mentioned that once again things seemed to be out of her hands, and she might be forced to quit her job after all. She was terrified again, that it might be cancerous, and she had decided to take a few weeks off from work and go in for further testing.

The tests came back negative. It turned out to benign and small enough, and growing slowly enough, that they decided against surgery or other extreme measures. It was such a disturbing experience after having the one with her house that she finally concluded that life could be short, and it was not worth it to spend the rest of her days until retirement at her abusive job. Besides, her boss had become completely fed up with her and was threatening to fire her. At last, she went searching for another job and found a sales job. She left her old job, but she was not happy with the sales work even though she loved the people she worked with. She finally decided to talk to some people who owned a gelato store and was determined to learn the business and put some of her savings toward her own business.

When she told her daughter, her daughter was fully behind her and assured her that she would be fine with finishing college based on the way things were going. This client's depression remained lifted, she was still off medications, her anxiety was significantly diminished, and she was sleeping well again. Nevertheless, she still admitted she was very worried about her future, but also, for the first time, had real hope that she could make it all work. That was the last I saw of her and so we closed the file on her case. I like to think it all worked out according to her dream, but it may have worked out differently but even better.

In these and similar circumstances, including family crises, brainwave training was confounded or shaped to a significant extent by unseen factors, until the source of trauma was uncovered or the client was persuaded by their own insight to take appropriate action that they feared and had avoided confronting. The young boy had to be removed from the school environment of an abusive teacher, and the single mother had to confront her acceptance of abuse rather than face her fear of homelessness grounded in her childhood abandonment issues. Judith Lubar (Lubar & Lubar, 1999) was among the first to document this effect at Joel Lubar's research clinic in Knoxville, Tennessee where he performed his groundbreaking research on attention problems and the type of solution NFB could provide. Her records clearly showed deficits in the neurofeedback training records that coincide with family trauma. These cases frequently involved events such as the death of a beloved grandparent, or abusive or alcoholic behavior by a parent. In each case, the child improved significantly from the NFB training, and then suddenly failed to make any further progress. The regression in behavior occurred at the same time as the precipitating event and it could be readily identified in the training session records.

We have found that professionals, such as chiropractors, who are not trained in psychology, are often blind to this type of confound and typically overlook it as a source of diminished efficacy with respect to neurofeedback. This may result in an endless search for better protocols or more recently developed exotic equipment to "fix" the problem without due regard for the real source of the problem. By the same token, we have found that psychologists and counselors frequently fail to recognize the impact of physiological and metabolic constraints on the brain and their impact on mental health.

In response to the need to raise awareness regarding the impact of socio-emotional variables on mental health and neurofeedback, I developed the Interactive Self Inventory to measure key dimensions of socio-emotional behavior. This assessment links the psychological dimensions of anxiety and depression to a variety of social dimensions of behavior to help the clinician better understand the source of the clients' behavioral issues and determine how much of their difficulty is from socially maladaptive behavior learned in a family system impacted by toxic stress and leading to complex trauma. Determining what effect complex traumas may be having on their health versus purely organic causes is a major challenge in clinical work. In fact, the ACE study shows us toxic stress often leads to both psychological and physical health problems. The ISI helps distinguish which is playing the heaviest role in a particular client's situation. I will review this instrument in depth in a later chapter.

Emotional Trauma and Brain Development

As mentioned earlier, there is an extensive peer reviewed literature on the impact of physical trauma upon the brain, but in the last decade there has emerged a fast-growing mass of evidence showing emotional trauma can also deeply impact brain structure as well as function. Volumetric studies of emotionally abused children indicate significant reductions in brain mass develop over time. Changes in the structure of the cortex and limbic system have also been documented. This leads to changes in the EEG distribution as well, but emotional trauma often has a more complex impact in terms of its brainwave expression than does TBI or multiple subconcussive events. Physical trauma to the head usually results in localized elevated eyes closed delta with generalized slowing and disturbances in coherence. Emotional trauma impacts microbiome populations and consequently can result in enteric inflammation. We have discussed recent studies that show this in turn leads to exchanges between enteric and cortical structures leading to pro-inflammatory responses involving the kynurenic pathway (O'Mahoney et al., 2014). This pro-inflammatory response also results in deficits in serotonin. The overall outcome in terms of EEG is usually in the form of elevated alpha that is slower than normal in frequency, along with significant alpha and beta asymmetries. So as can be seen, various forms of trauma can be distinguished in the qEEG as well as by the ISI and other assessments.

Emotional trauma also impacts the brain in another indirect fashion. This is through oxidative stress vector and the resulting excitotoxicity. Research on trauma shows individuals who live constantly in fear have chronically elevated cortisol levels and experience significant sleep disruptions. The constant fear generates excessive activity in the amygdala, and the brain responds to this excessive fear input through cortical activation in the form of the constant activity of worry and rumination. This activity, as mentioned previously, helps reduce fear inputs to the cortex from the amygdala. The worry generates chronic overactivation of the cortex resulting in excessive buildup of a neurotransmitter/metabolite known as glutamate. Sleep, it turns out, is one of the methods by which the brain clears itself of metabolites and byproducts of inflammation (Natale et al. 2021). The buildup of excessive glutamate and the reduction of pathways to reduce its presence results in a toxic environment for neurons that frequently results in neuronal decay or apoptosis or death.

It is not hard to conclude from the foregoing that emotional stress leads to functional and sometimes organic losses in the brain. The two processes are bi-directional and difficult to tease out. Sleep, along with exercise and good nutrition in addition to recreational activities are key elements of reducing the effects of emotional stress.

Metabolic Limitations

Over the past decade I have presented workshops on the "metabolic limitations" that can confound neurofeedback. My first official workshop at ISNR on the topic in the first decade of the 21st century had only eight skeptical attendees. The prevailing bias was that NFB could fix everything since it was targeting the CNS which controlled the whole body. This simplistic view lasted only as long as a clinician was willing to bury his or her head in the sand and ignore research from other fields. My most recent workshop, in contrast, had almost 40 very interested attendees, which for ISNR is above average, and a DVD that is selling well enough that I still get emails about it. This reflects the slow but growing awareness in both the clinical and public domain of the importance of nutrition and other factors that influence brain structure and function. It also reflects the growing awareness among neurofeedback clinicians that there are many physical disorders that profoundly affect mental functioning and limit the impact of neurofeedback because the brain does not have the appropriate resources to respond to training.

A prime example of this conundrum is hypothyroid function. It is a good example because it is so common at present as one in five women have it, and the majority are undiagnosed (Chiovato et al., 2019). It has a clear qEEG signature, and it results in symptoms similar to depression. In fact, there are many counselors and psychologists unknowingly thinking they are dealing with depression that is psychological in origin and trying to overcome it through talk therapy, little realizing that depression is physiological in nature. To confound the problem even more, it is well-recognized in research circles that the present test for thyroid function is very inaccurate, which is also covered up or ignored by the medical community. I routinely find the correlates of thyroid dysfunction in qEEGs and recommend a visit to a specialist only to be firmly told that the client had already been tested by their doctor and been given a clean bill of health. The ones I can convince to go to a specialist are always very surprised to find they do have a modest deficiency and pleased with the changes that treatment brings. The ones that do not go either need to be convinced in the course of training or sent on their way because they cannot physiologically respond to training. The brainwave patterns of those with thyroid issues are very resistant to NFB training.

The thyroid is one of many important indicators of developing inflammation from leaky gut. The inflammatory response is epidemic in our population at present. Both adrenal and thyroid function are confounded by developing inflammation. Their lines of communication with the rest of the metabolic system are disrupted by it. The consequence is a dysregulation resulting in a decrease in the production of thyroid T3 which is critical to neurodevelopment and function. Thyroid T3 is produced by a natural metabolic reduction of thyroid T4, which is typically what is tested for in acute care medical practices. The tests for T4 levels are very insensitive and consequently inaccurate (Dietzel et al., 2012). The test for T3 is rarely carried out but starting to draw more medical attention. Deficits in these hormones have been closely tied to exposure to pesticides by the neurotoxicology research community, especially at levels found in the food supply (Gilbert & Zoeller, 2010). The resulting deficits in T3 in pregnant mothers can result in profound effects on neurodevelopment of the fetus with respect to structural differentiation and growth of neural tissues. In adults, the reduction of T3 and T3 transport mechanisms overseen by astrocytes in the brain results in a deficit at the neuronal level. T3 is critical for DNA synthesis and enzyme production in neurons. A deficit in this cascade results in diminished output that appears as significantly elevated and slowed alpha in the qEEG and in the raw EEG morphology as well (Neidermeyer & Lopes da Silva, 2005).

The symptoms of thyroid deficits are sleep disruption, low motivation, mental fogginess, moodiness, low energy, and irritability. Thyroid dysregulation contributes to reduced myelination and a decrease in voltage-gated sodium current density reducing conduction velocities (Dietzel et al., 2012). Changes in speech, hearing, smell, and memory are consequences of latencies in visual auditory and somatosensory processes that have been documented, along with the general slowing of the alpha rhythm. It is interesting to note, as a major trainer in the field, how many clinicians I have encountered who are trying to train away this basic metabolic problem with operant conditioning of the EEG. It cannot work because the body does not have the resources to respond to training. It is like asking a sedentary business executive to run the marathon.

Another key area of impact on mental function is enteric integrity. Various compounds irritate and inflame the intestinal walls resulting in opening between cells in the intestinal walls referred to as tight junctions (Edelblum & Turner, 2009). This "leaky gut" syndrome results in the passage of large proteins into the blood stream which results in a strong cytokine response from the immune system. Over time this can result in considerable inflammation throughout the body affecting general organ function, and burdening the liver with detoxification efforts involving excessive lipopolysaccharides and other toxins. In addition, a breakdown in the flora population of the intestines results in reductions in amino acid synthesis, as well as affecting enteric serotonergic signaling, and this in turn affects the CNS and the production of neurotransmitters critical for brain function (Mittal et al., 2017; Del Colle et al., 2020). In the long term, research shows this can lead to anxiety and depression that is non-psychogenic in nature (O'Mahoney et al., 2016). In these circumstances, neurofeedback alone is not likely to resolve anxiety or depression. This is typically a case of trauma due to environmental toxins found in foods. Food additives and preservatives such as BPA, trace pesticides, and GMOs as well as food container contamination can all contribute significantly to this problem.

Genetic polymorphisms also confound diagnosis and treatment by how they limit the molecular activity and stages of production of enzymes and proteins and other critical categories of biological resources. These polymorphisms, known as snips, result in a shortage of key methyl groups critical for enzyme production in the process known as methylation and in particular neurotransmitter production. One polymorphism that has an important impact on neurological function, in particular, is methylfolate, which affects B12 and B6 availability. A deficit in methylfolate can result in depression and consequently brain fog and difficulties with attention. Like thyroid inflammation resulting in T3 deficits, counseling or training with NFB will not clear this up as these do not address the root cause of the client's problem. It has become clear from the research that methylcarbylamine is a critical additional nutrient for those who have a genetic polymorphism that prevents them from access to a bioavailable form of B12 (Shelton et al., 2013). Therefore, this nutritional deficit then can be a confound to effective NFB as well since it would diminish an adaptive response to training and the symptoms would perhaps mislead a clinician's assessments before, during, and after training.

Another aspect of methylation is that it can affect the GABA system of many individuals. There are those that have difficulty with caffeine because they have a problem with adenosine production during the methylation cycle. Because of adenosine's influence on the arousal system and sleep, it plays a key role in several neurological disorders and diseases (Reichert et al., 2022). These individuals do not produce enough adenosine to manage their GABA levels (Yang et al., 2013). GABA is the key neurotransmitter for inhibitory activity in the brain and deficits in GABA result in higher-than-normal baseline metabolic levels, just like trauma. Individuals with this problem are easily over-aroused and have difficulty calming themselves after a stressful of exciting event. They are also more prone to anxiety. This can easily be confused with psychologically driven anxiety and the chronically elevated metabolic baselines of those with trauma.

As can be seen from these examples, a clinician should do a thorough evaluation of all aspects of a client's history when dealing with clients they suspect of having complex trauma. These factors can either emulate symptoms of trauma and mislead the diagnostic process, or they can confound the interpretation of responses to training by their synergistic activity with a traumatized neurophysiology. Too much caffeine in sensitive individuals can amplify their traumatically established baselines, unpredictably generating excessive anxiety and interfering with sleep. This could be incorrectly interpreted as an overreaction to training and result in an inappropriate change or discontinuation in a protocol. During AT training it is especially important to control these factors and not confuse the effects of emerging trauma with serendipitous consumption of too much caffeine.

Finally, we should include environmental toxins, especially water and air pollution, in this list of neurodegenerative substances. For example, the neurological effects of heavy metals from pollution are well documented and have been known for almost a century. Lead and mercury have a devastating effect on the brain generating high amplitude frontal delta known as FIRDA and many other heavy metals have a similar effect (Neidermeyer & Lopes da Silva, 2005). Excess aluminum can have a similar impact as heavy metals and unfortunately it is widely used in public water filtration systems. In a multi-center analysis conducted by New Mind, we found that a significant number majority of children and adults we found with slow wave activity also had high levels of aluminum in their hair analysis findings (Soutar, Hopson, & Longo, 2017). The list of various industrial pollutants and their impact on the human body and neurological system is long and beyond the scope of this book, but the outcome in the long run is a degradation of neural function that negatively impacts mental function and social participation at a productive level. It also creates physiological barriers to effective NFB, and needs to be taken into account when inspecting qEEG brainmaps and determining NFB protocols and other treatments in the clinical treatment plan.

These sources of trauma are the result of the physical initiation of trauma, as a consequence of social forces and how they shape industrial production and distribution of the resulting social resources. However, psychogenic trauma can impact intestinal function as well. In the initial stages of psychogenic trauma, reductions of psychological distress can alleviate the inflammation of the intestines and move the client back to a normative allostatic balance, but once a certain threshold has been crossed it becomes more difficult. The body, at a certain point in the process, becomes a repository of distress and begins to generate further distress in the system because the balance has been tipped too far in the wrong direction. As van der Kolk (2014) argues, "the body keeps the score." The psychological precipitates a physiological problem which in turn aggravates the psychological even further. A dangerous bi-directional feedback process begins that progresses in a destructive fashion and system entropy begins to prevail. This chronic phase is more complex and far reaching in its ramifications and requires a more comprehensive response than a unidimensional approach such as psychotherapy or neurofeedback alone. This is why we emphasize a bio-psycho-social approach which we presently refer to as functional NFB. Neurofeedback works extremely well with individuals using a multimodal approach to recovering health but is muted in its effect if the entire physiological system is too far out of balance.

Chapter 8

An Updated Integrated Theory From New Mind

A fair number of clients will not respond significantly to two-channel protocols targeting the anxiety that typically emerges after the first 15 or 20 sessions of neurofeedback. This is also around the time when the first pre-post maps are reviewed. The clients with this issue are typically backsliding both in their trend screen and in terms of their scores on the symptom tracker. As we have said previously, these are for the most part the clients suffering from significant emotional trauma resulting in PTSD or the more subtle version we know as complex trauma. This is usually the point in the training that we introduce alpha theta (AT) training in order to integrate that trauma.

Alpha theta training is also sometimes referred to as deep states training. The term deep states neurofeedback originally came out of workshop discussions with BrainMaster Technologies. Practitioners had kept referring to alpha theta training as neuromeditation, another term from the same source, and I felt it was important to make it clear that alpha theta training was technically not the same as meditation. In order to distinguish it from neuromeditation, I suggested we call the workshop deep states neurofeedback as I had done in my own recent workshops' series at the time. I was conducting a separate and different workshop entitled Neuromeditation which involved special training in alpha states, and I did not want practitioners to begin confusing the two subjects.

During the early part of the last decade, I had experienced a peculiar confounding problem for the area of alpha theta training. I had been doing workshops and lectures on alpha theta training for a few years at ISNR and AAPB conferences. Bill Scott was starting to do the same and had even shown up in one of my workshops at the Winter Brain Conference in Florida. We quickly developed a good friendship over our similar interests. We both had initially gotten involved in NFB as a consequence our interest in alpha theta training and done a lot of training on ourselves and others. There was a very specific screen design format we were all using based on Peniston's research as well as a whole clinical format for conducting a session.

Unfortunately, I began to see prefabricated training screen designs in various software for alpha theta neurofeedback that demonstrated a complete disregard for technical aspects of the training and reflected a profound ignorance regarding the method. Bill's response was disbelief that people were even using such screens. Many clinicians were consulting with me and telling me they were doing alpha theta training and getting hundreds of crossovers in a session but poor training results. Upon investigation, I found the screen designs were totally improper. What was worse, is that practitioners thought alpha theta was "just" a protocol or neurofeedback gimmick. The assumption was that anyone could just do it if they picked the right protocol from a list and just ran it on a client. This indicated there was a profound confusion and ignorance over what alpha theta training was and where it came from. There were even many trainers, some of them very prominent, talking about AT who never used it clinically but had only taken a workshop in it themselves. In response to this dilemma, I called up Bill and proposed that we begin an effort to generate a series of workshops together to help better train the clinicians in the field and raise consciousness regarding the methodology. We set up workshops in the Netherlands and at BrainMaster in Cleveland, Ohio and began an effort to clue people in.

A Psychological Theory for AT

Not everyone doing AT work is a trained therapist. Many argue that only a trained therapist should engage in this kind of work. There is some merit to this argument if you are dealing with serious trauma, but it is different when dealing the typical level of trauma encountered in socialization process. Biofeedback is technically a form of self-regulation, and in that sense, more akin to yoga. There is an innate ability within all of us to heal from trauma whether it is physical or psychological. For the most part, AT is truly a testimony to the ability of each person to heal him or herself when provided with the right resources. It is from the AT experience that the therapist or the life coach comes to an understanding of the AT process and what to expect from it. Like so many other things in life, you must learn it firsthand through experience to effectively teach others. This is consistent with the experience of every clinician in AT that the administering agent of AT must experience AT personally first in order to be truly effective in their work.

In an interview that Bill Scott and I did with Nancy White at the Winter

Brain Conference (2000) she explained that she never processed with clients after a session but only had them fill out a form in the same manner as Peniston. This being the case, and it was a common method, then why is a degree in counseling or psychology necessary? If a degree in counseling or psychology is necessary, then why should physicians be allowed by license to engage in psychology? The truth is, that AT is more akin to working in hypnosis but even less directed. This does not mean that in the hands of an experienced and well-trained therapist, additional methods could not significantly enhance the AT experience, but it does argue that AT does not require this type of knowledge unless you are using it to resolve complex trauma or PTSD. Those individuals who have significant trauma require clinicians specifically trained to manage the process because abreactions can result with episodes of suicidal ideation and brief episodes of intense grief or anxiety. As Peniston noted, this is part of the process of healing, especially with AT. So, there is a bifurcation with regards to the application of the protocol based on the individual's severity of trauma. Trauma lies on a spectrum from the average amount experienced in the developmental and socialization process to the more moderate and severe cases that require special knowledge and experience to handle. The typical average cases can be managed by well-trained life coaches properly trained in NFB but serious trauma requires professionals trained in how to manage trauma. The first path is one of peak performance and the second is one of therapy.

It is obvious to every good clinician, and it is well known in counseling that some basic clinical skills are necessary to do healing work with the psyche. They are often the same skills called upon in the medical profession or pastoral work. Unconditional positive regard (Carl Rogers, 1961) and a compassionate perspective is the most fundamental feature of the healing environment. One should not enter AT work without these basic abilities. Unconditional positive regard, compassion, and active mindful listening are common features of most basic therapeutic methods. These dispositions are essential features of the Christian faith and its approach to healing as well as the part of the core principles taught by Buddha, or the Tao as taught by Lao Tsu. Both east and west share this fundamental criterion for the healing experience. Those who are not deeply motivated by this humanistic perspective would do us all a favor by staying out of this theater of activity. AT training grew out of a scientific transpersonal perspective that acknowledged the importance of qualia and consciousness and the role they play in the human experience. This is a holistic integrated paradigm of 21st century science grounded more in the monistic idealism of quantum mechanics proposed by Goswami (1993) than the 20th century quantitative reductionistic perspective.

In the Hero's Journey toward redemption, as described by Joseph Campbell, healing and transcendence are universal and fundamental themes, and there are pathways and milestones that can be identified and utilized to enhance and promote the expedition. Those who have a map are less likely to wander lost in the wilderness and more likely to find their "pearl of great worth" more quickly. This is not the realm where the credentialed authority of bureaucratic expertise in policy and procedure reigns—this is the realm where individuals with such agendas falter and fall. It is the realm where the unexpected and the unknown guide the way, and a realm where the Hayoka warrior rules by the mastery of letting go. The map required is a guide that links the collective unconscious to the present reality such as proposed by Jung. AT is the bridge. The journey for each explorer is infinitely unique. The ultimate authority is the individual's own inner guide. No person can take responsibility for the outcome other than the client him/her self. To declare any authority in the process is to undermine the client's process and thereby to possibly "do harm." Clients not willing to accept this are not worthy candidates and consequently will be judged and rejected by their own hand. The "therapist" is not doing anything to the "patient" and fixing cannot be part of the relationship. Therefore, this process ultimately lies outside of the authority of the clinician and the state. AT is not and cannot be psychotherapy. It is more than that. The individual guiding the AT session cannot entirely operate entirely from the traditional therapeutic model. I will come back to this theme in a later chapter.

With this in mind I will present a basic schema or framework for approaching this expedition into the unconscious mind. This is based on the literature in psychology, anthropology, and sociology as well as personal training in shamanistic traditions.

Adam Crane's (Crane & Soutar, 1999) theories of negative rehearsal and trances derived from Wolinsky (1991) fall along the same lines. My own theories outlined in the Automatic Self (Soutar, 2015) include all the foregoing, and are derived from the integration of a large number of theoretical paradigms as well. Those engaged in this "Process," as Crane referred to it, often emerge with very similar perspectives from their clinical experience. Adam and I explored these similar ideas that we also discovered in our parallel experiences with AT training. So, there appears to be considerable independent agreement emerging from different experiences in time with this process of AT. I will expand and refine these ideas in the section on Trauma.

Liminal States and Arousal

Arousal theory in psychology, sleep research, and meditational research all carry considerable consensus on this topic of arousal and liminal states. A good place to start with on this topic is to establish a basic schema of EEG patterns as they relate to arousal and the descent into sleep. Barry Sterman was originally involved in sleep research and wrote an article on arousal in the brain (Sterman, 1996) directed at the NFB community. In it he clearly outlines the EEG patterns associated with arousal and the entrance into sleep. The typical normative pattern for entering into sleep is not the same for those with mental disorders or trauma because they do not start from a normal distributional pattern of EEG frequencies in the brain.

The eyes closed EEG of a normal waking adult has alpha as the highest amplitude frequency, then theta, then delta and then beta. When the ratio between these component bands is correct, then a brainmap will show everything in the normal range. The abnormal pattern, on the other hand, can show up in almost any imaginable way.

When individuals first become drowsy there is a gradual increase in alpha amplitude and alpha frequency and then the alpha frequency begins to slow as shown in the diagram below.

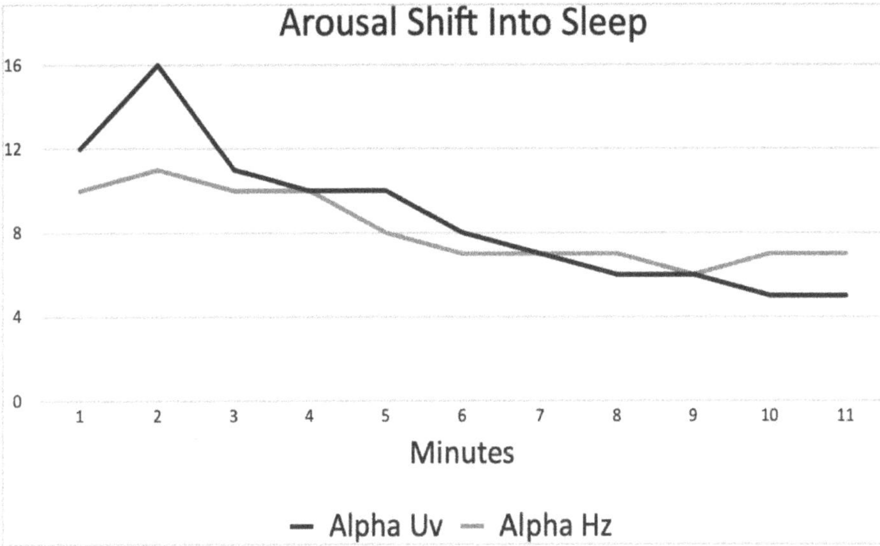

As drowsiness increases the alpha begins to drop in amplitude as shown by the black line. Alpha frequency also begins to decline as shown by the grey line. The two begin a gradual descent together but at different rates depending on the individual and alpha continues to drop in frequency and in amplitude with further increases in drowsiness. After a significant decline of several microvolts, theta gradually begins to increase until it becomes higher than alpha and the two converge and crossover each other. This is the crossover indicating the liminal state we are seeking where hypnogogic images and integration take place.

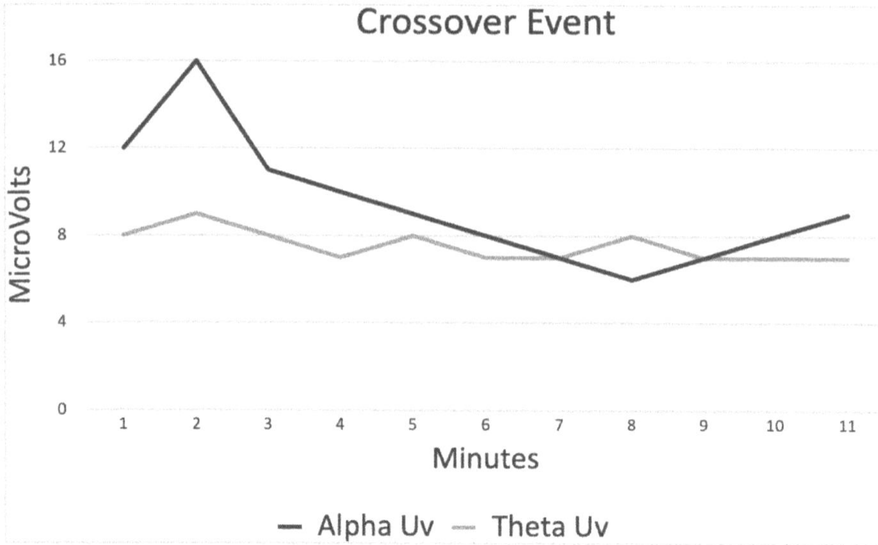

At the point where alpha and theta cross, delta will typically increase as well as the individual is approaching stage 1 sleep. To his credit, Egner (Egner et al., 2002) reported the AT details many others have missed. He notes alpha slowly subsides and theta and delta become more prominent (Broughton & Hasan, 1995). The phase when those frequencies become more dominant than alpha is usually when reductions in consciousness and phase one sleep occurs. AT teaches the subject to be conscious when they normally would not be by approaching those states that emerge from the slowing of the EEG but not entirely succumbing to them. He expected that both within and across session theta to alpha ratios would increase and they did not. He notes this makes the crossover states at times difficult to distinguish from a sleep, and indeed many people enter a sleep like state, but it does not have all the typical features of sleep. He comments that Moore et al., (2000) make a similar argument and that AT ratios did not differ significantly from alpha only training or EMG training.

If delta suddenly increases significantly in amplitude as in the figure below, the individual will enter a stage 1 sleep-like state. Note the precipitous increase in the black delta line the minute alpha and theta crossover in the trend screen below. The y axis is in microvolts with a scale of 0–20 μV and thc x axis is in minutes ranging from 0–30 min. In this case the individual has fallen into a range that would normally indicate deep delta sleep, but their brain is still responding to the audio input (Andrillon et al., 2016). So, they are still doing neurofeedback, they are just not conscious of the fact.

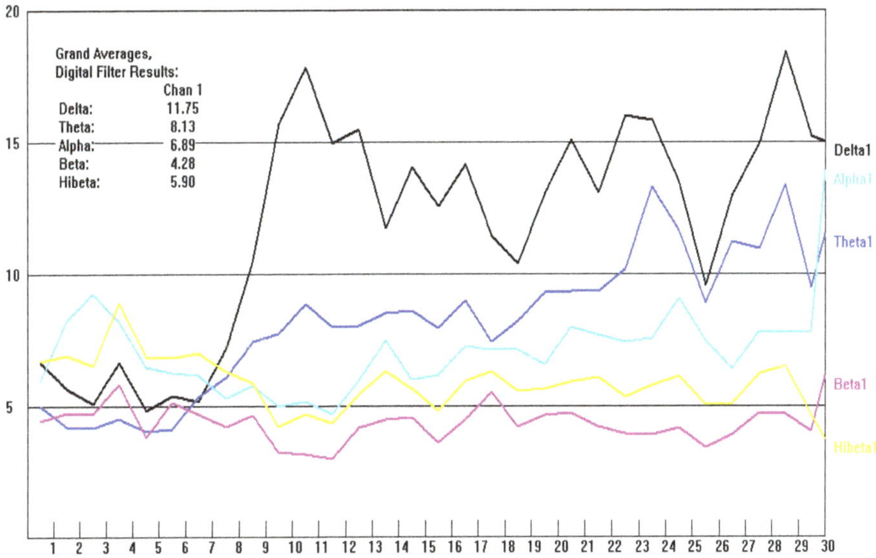

When the crossover occurs it typically lasts for a few minutes at most but at times can go on for 10 minutes or more. As the research indicated in the previous chapters, the more frequent and shorter crossovers are more desirable and appear to be correlated with faster and more significant progress. This is likely because the individual does not lose as much consciousness on average when shifting through the states associated with these frequencies. The other important feature of the crossover is when the alpha again rises above the theta, which is the point at which the individual becomes conscious of what insight or images have occurred. This was what was referred to as "paradoxical alpha" by Elmer Green. The software encourages the brain to keep oscillating between semi-conscious and conscious states.

This realm between the first increase in alpha amplitude and stage 1 sleep is not extensively researched and is poorly understood. There appears to be several possible states of consciousness in between that typically we spend little time in because of our cultural conditioning. For us this is just the corridor to sleep; but this corridor apparently has several doorways we never enter. In a sense, we all walk around believing we live in a two-story house of consciousness with the top floor of "awake" and conscious and the bottom floor of "asleep" and unconscious. In actuality, it is more like a high rise with a dozen or more floors in between the conscious and unconscious states but we always get into our consciousness elevator every night and push the "Lobby" button. We take the express ride down (sharp rise in delta) and never top at the floors in between. They are hidden from us by our cultural conditioning. It is interesting to note that psychedelics also puts us into these slow alpha states but they boost our arousal at the same time, so we are in a waking dream-like state.

This corridor we pass through into sleep therefore contains unusual liminal or "in-between" states that can be developed and enhanced. In the research on meditation the data shows consistently that advanced meditators using traditional concentrative meditation techniques develop over time the ability to occupy this zone at various levels at will (Dunn et al., 1999; Banquet, 1973). The initial work of Maxwell Cade (Cade & Coxhead, 1987) found that the more advanced the meditator, the lower the frequency of the alpha they would tend to occupy. Beginners often appeared in the 11 Hz range while the intermediates would occupy lower 9 Hz ranges and advanced 8 Hz or even 7 Hz. The extensive and replicated work of Banquet (1973) also has similar findings. In addition, Cade found the hemispheres balanced in power rather than asymmetrical and Banquet found high levels of coherence. The Japanese investigators Kasamatsu and Hirai (1969) who examined beginning and advanced Zen meditators at a monastery found the same pattern. More recent research indicates alpha production is tied in part to a type of somatosensory attention found in meditation that enhances awareness of attentional drift, allowing individuals to maintain state without distraction (Kerr et al., 2011; 2013). The difference between crossover and meditation is that alpha remains higher in amplitude than normal rather than dropping in amplitude. This is more extensively discussed in the chapter on meditation.

Typically, meditators dwell in these states for considerable periods of time ranging from 20 minutes to an hour on any given session. In alpha theta (AT) training clients zip through these states into stage 1 sleep and then bounce out into the paradoxical alpha state before repeating the shift again. This paradoxical alpha state is a period of brief consciousness which typically occurs in people before dropping off into deeper states of sleep. It is the period, as previously indicated, when people experience hypnogogic or hypnopompic images. Green harnessed this phenomenon so that he would sustain the paradoxical alpha response and use it, in a manner of speaking, to bounce individuals in and out of stage 1 sleep. Through this technique, he created the ability to briefly maintain an unstable but persistent occupation of this frequency domain without spending years developing the self-discipline, skill and brain structure typically required in meditation. It is a sort of "cheaters meditation."

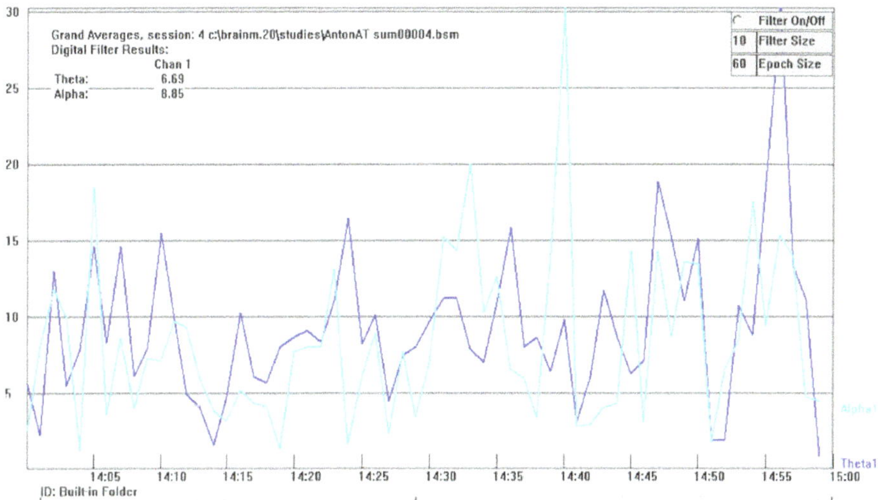

Alpha (light blue) emerges from its lower amplitude crossover state below theta (dark blue) 14:00–14:30 and paradoxically rebounds 14:30–14:45 before stage 1 sleep.

In order to sustain the levels of consciousness involved in meditation, the brain has to actually grow new networks and change. The work of Richard Davidson et al. (2003) shows that meditators develop various areas of the cortex that can be measured volumetrically as thicker. This revealed an entirely new dimension of the meditational process—it is not just a state that can be accessed at any time by an individual making the right effort, but it is a mode of consciousness that requires a specific brain structure to access and maintain. This is why long hours of practice over many years is required. This similar to the process of developing an athletic body and tuning for a particular sport. Alpha theta, on the other hand, does not involve this kind of sustained effort. It can provide a brief and albeit distorted glimpse of these states and the benefits they may offer, but it cannot provide permanent access. This is of course discouraging news to the typical American who believes it is his natural birthright to have anything he wants the minute he puts down his money. This belief reflects the magic bullet theory of everything which brought us the pharmaceutical epidemic (Whitaker, 2010). We expect for every problem there is just the right pill to quickly fix it if only we can find it. It also explains why some people have these wonderful experiences with alpha theta that are hard to repeat or why they cannot easily return to these states once they leave the clinic. I have had several clients who spent ten thousand dollars or more on special NFB sessions with a renowned expert in generating wonderous states of awareness and then found they could not sustain them. They found to their bitter disappointment that they had to engage in sustained practice and lifestyle changes to permanently access these states.

This does not in any way invalidate AT or its effects. For many they are very real and have an impact that is permanent. The effects are permanent because the client still does the hard emotional work of integration and learning and then adjusting their life accordingly. With AT, they can just do it more effectively and rapidly. To just enter these states for a brief period of time is transformative for many people. If they have already spent considerable time doing counseling and or inner work, it may provide a tipping point that does lead to permanent access to new levels of awareness. If they have done no inner work, they may experience some modest changes. If they have experienced significant trauma, on the other hand, they are very likely to have powerful experiences recovering lost sensory and emotional information related to the trauma as well as significant relief from its integration. This may feel like a dramatic change in consciousness, but it is just a return to a healthy level of consciousness and self-awareness. It is profound relief from sustained suffering.

Rescripting

Tom Budzynski introduced to our field the idea of "rescripting" the unconscious using various means of visualization and audio programs, often in conjunction with photic entrainment (Budzynski, 1999). In considering the concept, I would argue that Norris had employed visualization or guided imagery as a form of rescripting as well. It was also utilized by Green in his early work with biofeedback in conjunction with visualization, and it was natural for this nomenclature to migrate into neurofeedback and AT training through the community meetings. Peniston had already introduced the idea of the abstinence script based on his workshop experience with AT, and this work of Budzynski was just refining and extending the concept into other areas. Adam Crane called it "mental rehearsal" and many professional coaches were starting to use it in sports in the 80s and 90s. The idea is as old as shamanism (Noll, 1985). Visualization is clearly a more primitive aspect of this rescripting practice which is more basic and fundamental because it does not use verbal language. This idea is also related to post hypnotic suggestion techniques. In all of these techniques, an alternative vision of behavior or life in general is constructed and planted into the unconscious mind (think default mode network) in the form of a request or command.

Most authors favor the request approach and the only one to use the command version I have encountered is Peniston, although Green may have believed that the command approach was valid, as mentioned previously, when considering the influences that shaped his perspective.

By presenting in repeated fashion an alternative version of circumstances to the unconscious it is expected that it can be programmed or changed to act differently. This technique has proven very effective in sports and is still used today to enhance performance. I used it very effectively with AT training on professional golfers for years with excellent results. The shamanic version, however, is more detailed, symbolically integrated with the individual's unique unconsciousness aspect and it tends to be more global and profound in its results.

State Dependent Memory Theory

The theory that memories were tied to states of arousal or states of consciousness emerged in the latter part of the last century from the field of psychology. A large body of experimental work supported this idea, and it became common knowledge and is presently part of every undergraduate psychology curriculum. The classic example is the alcoholic who hides his bottle when he is drunk and cannot find it when he is sober, but remembers again where he put it then next time he becomes drunk. This applies to more subtle states and emotions as well. We may remember sad events better when we are sad and happy events more when we are happy. In fact, much of memory is indexed by emotion. Both emotional valancing and semantic memory networks converge in the temporal lobes where limbic process and cortical process meet. Close by is the insula that also mediates the intersection of limbic and somatic activity through what is known as the parahippocampal network. Combined, these structures and associated networks work to manage the emotion-memory complex.

Guiding clients into liminal states or providing access to deeper and calmer states of consciousness often opens the door to feeling forgotten emotional states, and the associated levels of arousal that result in the recall of memories not typically available to the individual. One of the neurophysiological mechanisms that leads to the loss of access to certain negative feelings and emotions and their associated sensory data that constitutes a memory is described by the work of Joseph LeDoux (1996; 2015). LeDoux found that a limbic structure known as the amygdala specifically records and indexes basic sensory data associated with fear. For most humans this process is applied to any internal or external events and memories associated with this emotion. When this structure becomes overstimulated through excessive trauma it may become overactive and dysregulated. This results in chronically elevated central nervous system (CNS) activity with its associated negative metabolic consequences, that van der Kolk (2014) refers to in his thesis of The Body Keeps the Score. The main strategy of the human cortex to protect the CNS from experiencing too much of this chronic condition is to increase activation of the orbital frontal lobes where the majority of the outputs from the amygdala terminate. This activation comes in the form of worry. In EEG terms this manifests as elevated frontal beta (Engels et al., 2007) and beta asymmetry (Avram et al., 2010) with excess beta in the right hemisphere.

Worry, like any process that gets repeated over and over again in the brain until it becomes routinized, creates structural changes through long term potentiation (LTP). The brain grows new pathways to maintain the activity and then places it into an automatic mode of operation, or at minimum, a reflexive mode that continues below the threshold of consciousness, so that attention and other resources can be diverted to more important immediate concerns. Since the majority of our thought processes occur below the threshold of consciousness in any case (Koch, 2004), this is a natural and easy procedure for the cortex to execute.

Consequently, large amounts of memories and their affiliated emotions become blocked and stored by the brain for processing at some later date—which never occurs if the new danger to worry about is the awareness of the memory of the trauma itself. In other words, the brain gets stuck in a catch-22. This negative feedback loop prevents the review and resolution of emotional trauma or PTSD.

The emersion of the individual into passive states of awareness reduces worry and allows them to access the issues and events that are the source of the fear that elicits the worry, providing the setting for integration. Memories emerge that previously were fragmented and not integrated into the conscious narrative of the individual. As previously mentioned, research on memories tells us that memories are not stored as one consolidated record but are recreated in real time, as the individual engages in the memory, drawing on discrete components from each sensory system. This requires the individual's memory system to draw upon these components and coordinate them, integrating them into one consolidated memory process. This process then emerges, as LeDoux indicates, into the conscious narrative of the individual. Memories are, in this manner, derived from an implicit subcortical memory system and integrated into an explicit cortical memory system. The implicit system is therefore a non-conscious processing system, and the explicit system is the conscious integration system that makes the memory part of the personal narrative. The binding of the memory is a crucial but little understood mechanism and is central to the entire binding process of the brain which coordinates all activity at every level.

One of the key difficulties of binding a memory due to trauma is the reactivity of the amygdala. As memory fragments begin to merge the emotional valences in real time into a bound memory, they bring with them also built-in associated emotional networks. These networks co-exist within the subcortical structures in different frequency strata (Kirk & McKay, 2003). The intensity of the building emotional valences typically triggers the amygdala and related fear networks, and they in turn stimulate sympathetic pathways. This escalating condition is likely to bring the individual into a high level of agitation that borders on the threshold of panic, at which point the forming memory collapses into its components and fails to coalesce. This failure of memory formation is likely to repeat again and again and yet never complete. The resulting agitation is difficult to cope with and individuals are likely to develop adaptive responses to distract themselves. These can often be in the form of obsessive or compulsive behaviors (not OCD) as well as addictions in order to block feelings and shift moods. The AT state appears to provide a physiological context that is a "safe zone" for integration of traumatic memories, as Nancy White (1999) theorized. It may be a natural adaptation that evolved to allow us to integrate socio-emotional trauma and has likely been harnessed through a variety of techniques by healers going far back in time.

AT and the Default Mode Network and the Unconscious

The individual in the deep AT state is therefore engaged in the process of retrieving, re-assembling and then, through conscious review, integrating the lost memories. Joseph LeDoux (1996; 2002) describes this in terms of transfer from implicit to explicit memory. Memories, apparently, do not become complete and do not become integrated into the conscious self-narrative until this transfer takes place. This is how we construct meaning out of sensory information. Severe trauma may result in a failure to encode memories, or it may result in a failure to integrate memories. The events may be so emotionally disturbing that the organism refuses to integrate them to protect itself. Researchers have noted that when mammals encounter life threatening trauma, they shake to integrate the emotional component. If they are prevented from shaking, they may die. This is part of the parasympathetic response to intense sympathetic arousal. Such is the importance of emotional integration for a mammal's survival.

Humans both shake and cry. Without this response, they may become very unstable in psychological terms. Early psychiatry observed the importance of emotional integration and that it could result in peculiar responses when it was traumatically challenged in the form of dissociative disorders. It is common knowledge in psychology that individuals exposed to severe trauma sometimes lose the use of a limb or even become blind despite there being nothing anatomically wrong with those functions. Medical teams sent to Cambodia, after the Khmer Rouge, briefly and violently, took control of that country and then were defeated, discovered many individuals who had gone blind after witnessing extreme acts of violence toward their family members. In recent times, this form of PTSD has emerged in a large number of soldiers returning from wars in the Middle East. It is now understood that war has always produced casualties of this form; it often results in suicide.

Until recently, the mechanisms of this form of conversion have been little understood. This was partly because medical professionals did not fully understand the importance of emotional integration for mental and physical health. In large part, this was because of our Cartesian dualistic system of thought that was so cherished by western science, and the overfocus on material events being the most reliable reference for understanding solutions to problems. Healing became understood as primarily a physical event and the psychological would follow. In the early twentieth century medical labs became the basis for medical schools. Institutions such as John Hopkins secured their credibility and funding through these labs (Barry, 2018). Physical healing was a measurable process while the efforts of psychology labs, such as Wilhelm Wundt and Edward Titchener, were in doubt. Biology labs lent the prestige of physics to what was previously thought of as a mixture of craft and art. The placebo effect was discerned as merely a useful exception which would eventually yield to experiment rather than a flaw in the physical model. Healing, both physical and psychological, would ultimately prove to be best approached through the physical. So, drugs have credibility and AT does not.

The retrieval process may generate other related new details of memories as well. As previously mentioned, the body adjusts to both state and location with respect to memory, and memories emerge in these familiar intersections that are typically not available. The salesman who cannot remember the details of his contract until he arrives at the closing location is no different than the alcoholic who cannot find his hidden bottle until he is drunk again. Revisiting the theta frequencies that dominate childhood (Niedermeyer & Lopes da Silva, 2005) is also likely to provide memories of that period because the client is once again in that state of mind, especially if they begin to visualize the locations of childhood (White, 1999). During these episodes of deep states, many forgotten details emerge that begin to inform the adult mind with a new perspective, leading to profound insights into the nature of their childhood. These are the primitive and generating sources of their emotional scripts and their primal beliefs. These childhood memories may lay dormant, implicit, and unreviewed for decades, if not forever. The adult never returns to inspect the quality and validity of the assumptions and beliefs that form the foundations of their adult attitudes and behaviors. People act automatically based on emotional scripts forged in the formative years, and then as adults they rationalize and construct narratives justifying their emotional acts. This is the activity recorded in the default mode network research. This is the self process often referred to as ego.

Over the last 20 years, in clinics I have managed across the country, I have observed this process clinically, especially in individuals with dissociative disorder as they would recover memories in bits and pieces. As traumatic memories emerge, these individuals often take refuge by literally abandoning their conscious process. I have watched this occur while monitoring the EEG of these individuals and having them describe the experience afterward. They "just go away" and yet their eyes remain open and staring. Often, they will unconsciously move themselves to safer locations, such as a closet or under a table. We had one client that would literally drive away in this state unless we took their keys away before the sessions. Antonio Damasio (1999) notes that many of the severely disordered individuals in the institutions where he did his research would behave in similar manner when in unconscious dissociative states. He notes that the only detail that would tip him off that he was dealing with such individuals was the complete absence of emotion. He concludes that without emotional networks operating, there can be no consciousness.

The memories that do emerge from severe trauma do not typically emerge as one complete memory, and they do not require self-narrative activity to surface. The client tends to remember first a sensation or a smell, and then perhaps a voice or sound. These may seem to be happening in real time during their emergence and the clients may re-experience the touch and smell, even the voice, as if they are actually happening. This is the brain and the body attempting to access and re-sort the pieces of memory stored in different sensory regions and bind them into one coherent memory. These are old networks sorting themselves out in real time. This can happen repeatedly over a series of several sessions until the memory emerges as a whole. This is how lost memories really are recovered, not the pseudo lost memories often elicited in torrents of invented images that have been inadvertently coached out of patients in talk therapy.

This process of retrieval also relates to the synthesis of new ideas and insights and the flash of frontal midline theta that accompanies it. Without the material that is derived from their suffering, people frequently are unable to synthesize an authentic narrative of their life that provides meaning. Without this comprehensive narrative, the individual feels fragmented in terms of identity. The individual feels social and group support to authentically identify himself to others so they can determine what status he holds. When that status is determined then the consequential legitimate limits of his action can be determined. This dilemma is continuously problematic and must constantly be negotiated (Blumer, 1969). There is a dynamic associated with this politics of identity that drives networks to integrate, an implicit momentum to the process. This fragmentation generates a CNS imbalance that involves anxiety and sympathetic hyperarousal that is chronically uncomfortable and limits their abilities to integrate new material and self-soothe and nurture themselves. They often feel agitated and may suffer occasional panic attacks. Many discover early in life the use of alcohol or drugs to compensate and ease their discomfort and dis-ease. It is, however, the recall and reintegration of the lost memories and the insights that they require to heal. This is not at all self-evident until it begins to occur. Individuals who have not done this work on themselves, including therapists, are oblivious to the profound power of this tacit experience. This is not just a process of talking it out.

The new ideas and insights usually emerge spontaneously for the client. They may occur in the office or they may occur outside the office while the person is driving to work or taking a shower in the morning. Sometimes the AT session will generate hypnogogic images that jog the individual's memory of something in the past. Sometimes they will just spontaneously have a new insight. It is the resulting insight that flags their integration. It is the insight that reflects a new organization in the networks of self-organization. The outcome of the insight is likely to be novel ways of thinking, feeling, and doing. Spontaneous and creative interaction is restored. These novel patterns of interaction frequently lead to resolution of old patterns of behavior and the initiation of new responses and patterns of behavior that dissolve barriers to improved functioning in social contexts. The new organization of meaning is derived from a new configuration of the past that redefines the present and possible future. A missing piece that was felt in unsatisfactory experiences of social interaction, but not known in the cognitive domain, generates a new experiential gestalt.

This pattern of discovery and insight suggests the hypothesis that the integration of experience is crucial to maturation and development stages beyond childhood as Eric Erikson proposes. That we have another level of "knowing" than we utilize consciously, and that this level of knowing involves behaviors that elicit responses from others, full of emotional friction generating painful experiences, completely eludes us. We reside in the illusion that our conscious self is the determining factor in our behavior in the same way that we assume all learning is consciously mediated. This illusion is a stage of growth we eventually move through. It informs all the other stages of maturation and is part of a complement of forces that propel us toward a felt need to integrate with the world around us. As separate selves we experience a basic sense of incompleteness that social interaction resolves because we are social animals. Our ultimate identity derives from the collective.

This sense of incompleteness also has a biological component that goes back to birth. We learn to become separate against a background sense of loss of connection with our biological source and individually with our mothers. From this perspective we move out of the biological and into the social. We are always thirsty for integration at both the social and biological level and this is what motivates us and fulfills us. It is a force as basic and profound as gravity. The thirst itself becomes another level of inspiration that can drive individuals into addiction or spiritual pursuits, especially if they are frustrated in attaining social resources. Spiritual inspiration draws power from ongoing confirmatory experiences that neither social nor biological experience can permanently resolve. These dilemma has its root in the sense of separation that emerges from birth. We are always hungry. This is the energy that propels us through stereotypical stages of development that the boundaries of planetary life define for us.

AT, Synchronicity, and Insight

Carl Jung introduced the concept of synchronicity in the early twentieth century and he and the physicist Wolfgang Pauli began to work out a novel model of the universe combining psychology and the quantum perspective (Jung, 1973). Synchronicity proposes that the universe has non-causal mechanisms that generate physical coincidences tied to meaning and insight. Events in our lives are connected to our thoughts through meaning and they unfold in parallel vectors. In the same sense that mathematical laws parallel physical events, the unconscious process, and the archetypes it uses to process information also parallel these events. It becomes obvious when we become aware of having intense emotion and thought around a person and they suddenly call us on the phone. In this model, consciousness is a fundamental component of the universe that is irreducible like gravity. It plays a key role in the order of events through non-local effects due to the interconnectivity of all things. However contrary to our senses this seems, this is a realistic proposal that has a sound scientific basis in mathematical proofs of David Bohm's work on quantum holography and the implicate order (Bohm, 1980).

Odd as it may seem, synchronicity appears to be an unavoidable phenomenon in the healing process and especially with regard to AT. In my 20 years of AT work, synchronicity is so common it is difficult to ignore, and it can be operationalized to achieve enhanced outcomes. More often than not, it plays a key role in the fundamental elements of healing events leading to insight. Visualizations often intensify the process in AT. Visualization related to problem solutions also will result in the client discovering more novel opportunities and situations to solve their problem. More "physical" solution spaces will occur. Some of these situations will be a consequence of chains of events set in motion by different responses to social events and contexts. A portion of them will not be explainable from this perspective and will be relegated to the realm of coincidence. The number of coincidences and the specificity of their nature with respect to the client's issue, however, will typically be profound.

It does not appear to matter whether the client sees these as coincidence or not. It is important that the events be acknowledged as significant, and that clients review how they relate to a change in their attitude and orientation to their problem and possible solutions. As clients see themselves as capable of initiating novel responses and actions that result in solution-oriented outcomes, they begin to experience increased self-efficacy and reduced depression and anxiety. This in turn leads to more open perspectives with respect to future problems, and defining themselves as having multiple options and thereby expanding their perceived repertoire of solutions in general. This outcome results in a considerable sense of empowerment and positive affect. It moves their locus of control from external to a more internal orientation and consequently reduces their sense of powerlessness and frustration.

The Role of Dreams

The visualizations used in AT often migrate into dreams. As clients focus on a particular image, or archetype, it may occur in their dreams. In reviewing their dreams, powerful insights regarding the client's feelings can be derived. In addition, the location of the emotional drama in their life can be recovered by consciously reviewing where in their daily life they experience that emotion. This enhancement of self-awareness regarding emotion can also increase their general emotional self-awareness, as they begin to monitor their emotional responses more carefully to situations in order to index the emotions in their dreams. Clients begin to interpret their emotional lives with respect to their dreams and they begin to interpret the emotional content of their dreams with respect to their daily lives. The two become intertwined and integrated. This results in a dialogue between the conscious mind and the unconscious mind in a novel fashion. They experience an enhanced sense of integration and centeredness.

The emerging dialogue between the conscious and unconscious appears to enhance the synchronicity in the client's life, the greater the exchange between conscious and unconscious, the greater the level of synchronicity. This generates a sense of empowerment and, even more importantly, a sense of connectedness with the self and with the world through the implicate order. Clear themes emerge from the process. Events appear to line up with these themes. The individual establishes a very real internal connection with an older and seemingly wiser component that they had not realized was within them. This aspect of their self seems to have greater resources and appears to have influence upon the world at large. It is not misled by the confusing events of daily life and the razzle dazzle of the chattering rational mind. It reads the most intimate emotional core of the individual because it is part of the individual and yet it can provide commentary and guidance of an apparently transcendent character. It is this very wisdom that appears to guide the world forces around the individual.

In the effort to induce change through visualization, the individual will often find that the visualization themes they choose for AT training will appear in their dreams. Almost like a response or echo from the unconscious. A dialogue emerges between different conscious and unconscious aspects of the psyche. By picking up the themes of this symbolic thread and reusing them in the next visualization, a cycle of presentation and response can emerge that culminates in an actual physical event that appears synchronistic, as Jung would put it. It may be the appearance of an object or event that moves the individual in the direction of the visualized theme. For instance, visualizations about a distant relative or relationship will result in email or letters suddenly arriving from that individual which opens up new dialogue and insight to the relationship of concern. I had one client who had stopped speaking with her mother for many years and was very angry at how she was treated as an adolescent by her mother. After the first couple of visualizations, her mother sent her a letter asking forgiveness for how she had treated her and they ended up eventually having several productive calls that significantly healed the relationship. This allowed the client to make progress on her own personal relationship issues which were an important part of her goal. Synchronistic events such as this appear to be very commonplace in AT work.

It is interesting to note that one of the first things to go in mental disorder is the memory of dreams, and medications enhance this effect. Also of interest is the research that indicates that lack of sleep and dreams leads to progressive disorder (Walker, 2017). If you don't enter REM, you begin to hallucinate. Although the role of dreams in human life is still under ongoing investigation, it clearly has a more important role than previously entertained by the behavioral sciences. Nature does not waste time on useless processes. Arrogant scientists who rely too religiously on established trends have a long history of missing important insights into the world of nature. Scientific revolutions occur because of this narrow-minded focus that often dominates bureaucratic scientists (Kuhn, 1970). Being dismissive with regard to the phenomena of dreaming is progressively proving to be a foolish and naïve position. Lucid dreaming was dismissed as a phenomenon and consequently proven to exist through laboratory experiments and to even have a temporal dimension (Erlacher et al., 2014).

It has been observed for some time that as clients heal through the process of neurofeedback, their awareness of dreaming returns and they begin to remember their dreams. Evidence that dreaming is important for the consolidation of learning goes back to research in the 70s (Walker, 2017). There is strong evidence that dreamless deep delta sleep may be critical for physiological learning such as motor memory. The entire dreaming process appears to be related to consolidation and integration of learning. The theta frequencies in the 7 Hz range have also been found to be involved with the REM process, and midline theta is under ongoing investigation for its important role in waking memory processing (Klimesch, 1996; Kirk & McKay, 2003). I have proposed the theory for over two decades that dreams, and theta frequency images and emotions, are the basis for a fundamental mammalian process of thinking. The verbal overlay is just an artifact of our social nature as human beings. As it turns out, emotion and memory functions tend to occupy the same limbic networks at approximately different but parallel theta frequency bands (Kirk & McKay, 2003). This is discussed more extensively in NewMind course on qEEG (www.newmindacademy.com) and related electrophysiology. Theta also dominates the mind of children below the age of 5. As Nancy White (1999) notes, as discussed in previous sections, this information may be difficult to retrieve in normal waking states. Returning to the frequency they were formed in may assist and be part of the state dependent memory process. The most important point in all this is that considerable processing takes place in these frequencies before information becomes available in networks involving verbal processing and self-conscious networks in the frontal regions. The consequence is that considerable processing can be held up in these network domains because of emotional trauma.

The research on encoding failures and the failure of memories to bind when formed in situations involving significant trauma is becoming progressively more convincing (Samuelson, 2011). This apparently leads to progressive isolation of network groups that draw on similar sources for processing. This can reduce capacity and begin to generate conflicts in resource management that impedes physiological function. Freud became aware of the importance of dreams in the human psychological process as he began his investigations of the mind and worked with mental disorders. Jung made a career out of dreams. Dreams have a long history of being part of mental disorder and recovery. The Egyptians had the Dendara Temple in which priests occupied private rooms where individuals would come for dream analysis and guidance. The recovery of unintegrated information that was a consequence of trauma often emerges at least in part through the process of REM sleep and dreaming. Clinicians should be prepared to deal with this process.

Crossover Features Summarized

Elmer Green felt that crossovers were a critical component of AT because they often resulted in images providing solutions to problems posed in the AT state. The crossover state tends to produce vivid dream like images that flash before the individual's consciousness like an intense memory. They contain fragments of information that are often key to an individual's recovery and integration. They occur when an individual's alpha activity drops in amplitude and their slow wave activity, usually theta, increases. If too much slow wave increase occurs, the client slips into a state that resembles stage 1 sleep. For an individual with relatively normal eyes closed EEG distribution, the alpha drops and crosses over theta when looking at a trend or session review screen. The theta often increases a microvolt or so, while the delta tends to increase as well but not to the point typical in sleep. The theta then returns to a more normal amplitude and alpha begins to increase again generating "paradoxical alpha."

The dropout or sudden reduction of alpha amplitude, when it occurs, is rapid for most individuals. There is an initial decrease in alpha frequency from the normal range 9.5 to 10.5 into the 8 Hz range. It can even hoover below 8 Hz and may even contribute to the amplitude in the theta filter. Most individuals cannot remain conscious during this brief period of alpha frequency reduction and begin sliding into stage 1 type sleep rapidly. Practiced meditators, however, can maintain consciousness at this frequency range but begin to move into an altered state that is typical of meditation. This in fact is the region of neural functioning that AT targets, but most people can only remain in this region of functioning for brief periods and are only marginally conscious. Instead, the average person shifts in and out of this region closer to the border of stage 1 sleep, which is closer to a trance state, but they are in and out of self-awareness. Clinical experience indicates that the level of self-awareness is related to alpha amplitude, as Green surmised. So, he called the alpha that pushed people back out of stage 1 and into consciousness every time they became self-aware again "paradoxical alpha."

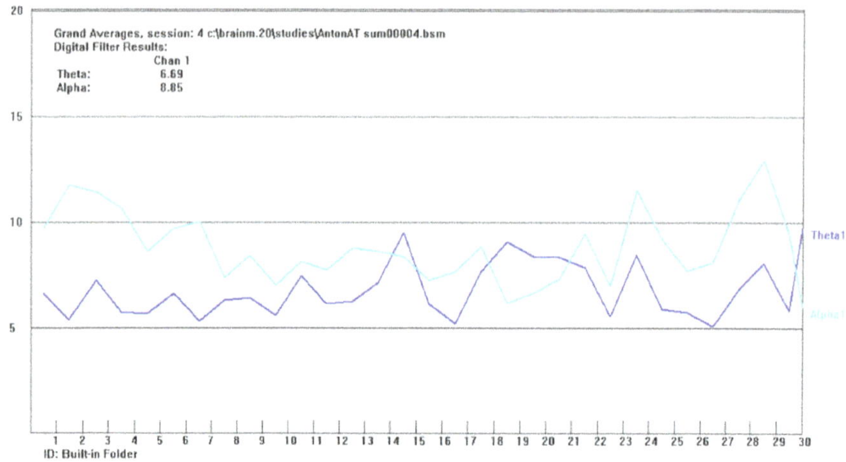

Minute of Session	Alpha Avg Freq
10	10Hz
11	10Hz
12	10Hz
13	9Hz
14	8Hz
15	8Hz

As we have mentioned in previous chapters, it is this shifting in and out of alpha while dipping into stage 1 sleep that generates the hypnogogic images. Not everyone has them (Moore & Trudeau, 1998), especially if they are having no dreams they can remember—which is a common feature of mental disorders. This may be tied to their EEG distribution and level of neuroplasticity in that their theta or alpha is chronically too high to begin with. What Elmer did not mention is that some people have hypnogogic smells, or sensations, or emotions instead. These are the fragments of sensory memories that have not been integrated into one memory and consolidated as a conscious explicit memory. These memories tend to be disconnected from their emotional components because they have never been fully formed and "tagged" with emotion by the hippocampus. This is part of the physiological process of memory formation and integration discussed by Joseph LeDoux (1996; 2015) that he uncovered in his research on the amygdala. Moore and Trudeau observe that clients may also just have visualizations or sudden insights that are just as critical to the integration and transformation process as the hypnogogic images. They may also have strong physical experiences and sensory events such as floating, falling, flying, local tension, or pain, etc. Moore and Trudeau noted crossovers can result in colors and geometric patterns, biographical memories, symbolic images, and heaven forbid, spiritual experiences.

As discussed in a previous section crossovers tend to breakdown as follows as follows: 27.9% have no crossovers, 34.1% have spontaneous imagery or memory events, 37.9% have spontaneous sensory experiences related to memory, biographic images occur about 7.1% of the time, perinatal images appear approximately 2.1%, and transpersonal images and experiences 6% of the time (Johnson & Bodenhamer-Davis, 2017). Shorter and more frequent crossovers in the range of 2–5 minutes tend to produce the best sessions and images.

Individuals with abnormal distributions like that shown in the figure below often have a very different pattern of "crossover" that initially confounds most therapists. For some people the layering of the EEG can be inverted with beta on top, then alpha, then theta and finally the beta frequencies. "How can they have a crossover if the alpha is already below the theta?" is the typical response of clinicians learning this process. For individuals with this pattern the same principles seem to apply but they appear differently. The alpha still drops out, though often not as dramatically and theta increases mildly while beta remains relatively the same. If they enter stage 1 sleep, then delta increases as well. In this case there is no official crossover to be observed, nevertheless the shift in consciousness is similar and hypnogogic images occur. In general, with most trend screen distributions of EEG, the alpha dropout with increased theta continues to define the AT crossover pattern.

EEG Morphology as the Brain Shifts Into Sleep

In the indicated trend screen below note the alpha is lower at the outset than delta, likely due to a TBI, and almost the same in amplitude as theta. There is a clear drop in alpha until around minute 9 when both theta increases slightly along with delta. This is the only indicator of a crossover, but the basic pattern holds regardless of the relationship or layering of the various component band frequencies on the trend screen at the outset.

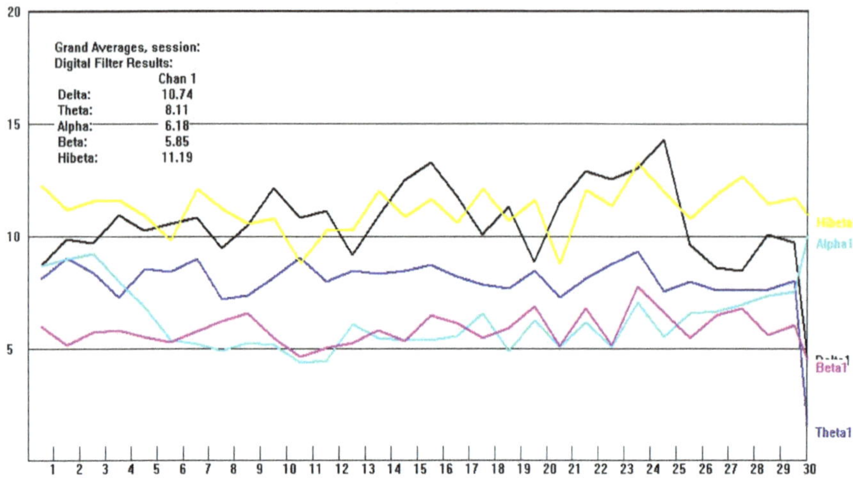

It is good to be clear on the standard definitions of sleep stages as it helps to assess the clients' state during a training session.

Stage 1

Stage 1 is the beginning of the sleep cycle and is a relatively light stage of sleep. Stage 1 can be considered a transition period between wakefulness and sleep. In Stage 1, the brain produces high amplitude theta waves, which are very slow brain waves. This period of sleep lasts only a brief time (around 5–10 minutes). If you wake someone during this stage, they might report that they weren't really asleep.

Stage 2

Stage 2 is the second stage of sleep and lasts for approximately 20 minutes. The brain begins to produce bursts of rapid, rhythmic brain wave activity known as sleep spindles. Body temperature starts to decrease and heart rate begins to slow.

Stage 3

Deep, slow brain waves known as delta waves begin to emerge during stage 3 sleep. Stage 3 is a transitional period between light sleep and a very deep sleep.

Visualizations can be as valuable as hypnogogic images in AT. Moore et al. (2000) investigated crossovers using percentage of time the two frequencies converged. This can be misleading, especially if averaging time is short and quick, swings in amplitude in alpha and theta are measured. What is of particular interest in this study is their observation that cross-over did not involve an increase in theta and a drop in alpha, but rather no change in theta and a drop of alpha below theta. They found this consistent with the literature on sleep and EEG and the phenomenon called "alpha drop out" (Niedermeyer & Lopes da Silva, 2005). This is also consistent with our clinical observations. In addition, as was mentioned, they found a significant number of subjects failed to have hypnogogic images but found themselves in spontaneous self-guided visualizations that were highly pertinent to their concerns. These images often lead to valuable therapeutic insights. Based on these findings it would seem a good strategy not to designate hypnogogic images as the only valuable feature of AT training with respect to crossovers.

A New Mind Standard AT Method

To begin with, hand warming and breathing exercises make it easier to produce both alpha and theta frequencies and this should be considered an important part of preparation for training. Today heart rate variability (HRV) is an excellent method for training breathing and reducing excess arousal. This is part of the stabilization training the ISTSS recommends prior to exposure work. So preliminary preparation is important. In that vein, fast wave training to stabilize and prepare clients for AT also aids them in dealing with trauma and recovered memories and helps them to be more present. A good session length is 30 minutes. Clients can train for an hour a day at home but for many every other day is a good pace that allows them to process whatever emerges from a session. Passive intent works better than active willful effort, especially with respect to the production of crossover imagery. Creativity often manifests in personal insights and solutions to problems through changes in perspective and behavior. This can often be spontaneous and is part of the integrative process. Mild abreactions are common in the form of fatigue and increased initial anxiety that eventually diminishes. Integrating trauma is challenging emotional work. More severe disorders like combat PTSD can result in intense abreactions that require significant clinical skill to manage. A reclining chair can aid some in accessing theta as well as dim lighting. The clinician does not have to be in the room and how much time you spend in the room depends on how much it aids or hinders the client's progress.

HRV Training

Heart rate variability (HRV) training has become the preferred mode of biofeedback in most of our affiliated clinics and a critical component of treatment. Biofeedback has always been an important part of the NFB training process but over time HRV has progressively demonstrated itself to be generally the most effective and user-friendly modality. One of its most important features is that clinicians can see both sympathetic and parasympathetic response in the feedback. We tested it against all other modalities and found it to be one of the most consistently effective across all clients. This is important because many biofeedback modalities are only effective with one subset of the population (Schwartz & Andrasik, 2015). One example of this is hand temperature training. Additionally, it is very user friendly. It only requires placing an ear clip on the client's ear lobe and involves no paste or special electrodes. Clients can purchase inexpensively their own version for home. Clinicians can take internet courses in how to optimally use their HRV and we teach basics in our New Mind Workshops. To learn about the research behind HRV please visit our New Mind YouTube site and view some of the many speaker videos we have on the topic from some of the best experts in the country.

Clinics utilize this modality of feedback in many ways. Our primary approach is to teach individuals how to breathe properly using the feedback from HRV. We learned that tidal CO2 is a crucial measure of proper breathing and HRV is an excellent proxy measure of this variable when breathing is done properly. The outbreath should be a little longer than the inbreath with some space between both to get the correct balance and many clients do well counting to themselves while breathing with 4 counts on the inhale and 5 counts on the exhale. Sometimes an added count after the inhale is important too. We have tested this with a capnometer device which measures the actual gas level in the nose. It is also important to emphasize diaphragmatic breathing and remind clients not to exaggerate their breathing. Most individuals with anxiety engage in chest breathing with short shallow breaths at a rate greater than 15 breaths a minute. When they are relaxing, we want them below 10 breathes per minute. Their heart rate is often in the 90s as well. We need to have men in the 70s and women in their 80s.

We typically train individuals for 10–15 minutes at the outset of each session. Then we begin the NFB session. At that point, we turn off the HRV feedback sound and remove the HRV video feedback screen from the training monitor. From that point on we continue to monitor their HRV while they are training to assess how well they have learned and generalized the training. We ask them to do homework and to practice, without equipment, several times a day for 5 minutes in order to help generalize the effect of the training. This works better and results in better compliance than asking them to do one extended practice. As the session progresses, we test them without feedback to see how well they can do it on their own.

This procedure has the effect of damping down the chronic hyperarousal that most individuals with disorders experience. It also helps to reduce their experience of agitation and anxiety when we lift their initial depression with the first 15 sessions of NFB. We believe that it also accelerates and enhances their NFB performance during regular NFB sessions.

This protocol is even more useful with alpha theta training. By the time we get to the AT training after 15 or 20 sessions of NFB, they are usually fairly accomplished at HRV. We can use this technology to relax them so they get an optimal AT session experience, but it is also an excellent indicator of how well they are doing with integrating trauma from crossovers, and provides notice of any trauma that may be emerging as they approach crossover. When clients begin to approach the crossover point, the HRV coherence rate drops significantly for those with complex trauma and toxic stress, indicating that traumatic memories, feelings, and sensations are starting to surface. HRV should typically improve as we move into sleep and not the other way around (Ginsberg, personal communication, 2017). As clients progressively experience more crossovers, their HRV coherence remains more constant moving into crossover and may even improve after coming out of crossover. This progression of improving HRV over successive crossovers empirically demonstrates successful integration as well as overall progress of the client. However, clinicians should be cautioned against expecting coherence to reach a point where it never degrades approaching crossover, or that it must improve after a crossover as a final expectation and measure of complete integration. As indicated by the previous literature review, the effects of complex trauma rarely fully resolve, even after 10 or 20 years of therapy.

Below is an actual review or trend screen of an AT session. It is a classic pattern in that alpha first increases and then begins its descent until it crosses over theta in the first half of the session, usually between 7 and 15 minutes. Alpha then increases in power as a consequence of the paradoxical alpha response and the individual often experiences a hypnogogic image. The alpha then begins another descent with multiple short crossover events that generate an appearance of alpha and theta intertwining and then another burst of paradoxical alpha. And so, the pattern continues. This is the pattern that eventually emerges after multiple sessions, and it may take 15 or 20 to achieve it. The first crossover may sometimes take seven or more sessions. The more trauma there is, the more likely crossover is delayed in its appearance.

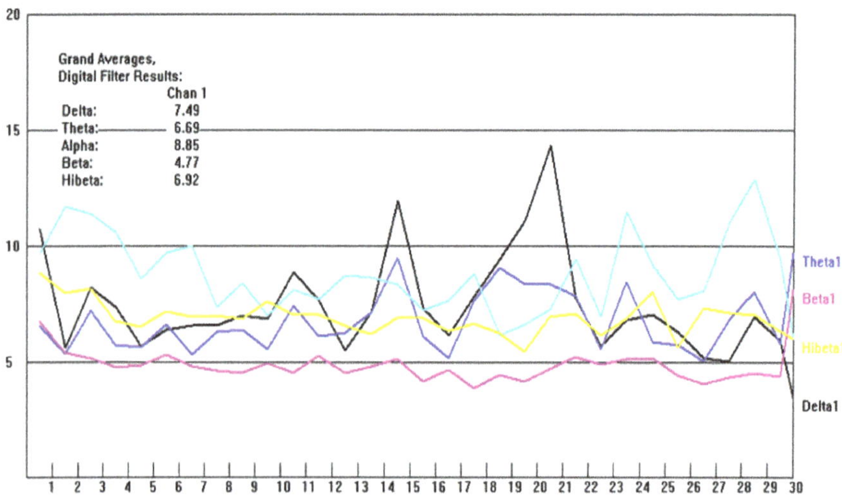

Crossover occurs when alpha (light blue) descends in amplitude until it drops below theta (dark blue) between minute 14 and minute 15 in this particular session (X-axis in microvolts, Y-axis in minutes).

The Design Screen

The figure below shows a generic design screen with the component bands displayed as bar graphs. Adam Crane built and distributed the first commercial NFB equipment designed exclusively for AT training. Bill Scott, however, was the primary researcher in the field and working with Eugene Peniston eventually defined the standard settings for a modern AT design screen. This design incorporated more than alpha and theta in its implementation. Delta and beta played a role as well. Bill found many individuals came into his office with excess beta because they had too much chronic hyperarousal of the central nervous system. Most clinicians at the time did not have access to qEEG and relied on observations such as this to guide their protocol selection. He set his beta filters to down-train the beta frequencies to help these individuals to initially calm down and then turned off the feedback on it once they achieved crossover. The alpha filters were set at 8 to 10 Hz to encourage the production of lower frequency alpha that would lead more easily to theta production. Theta was set at 5 to 7 Hz. Delta was set at 2–4 Hz to reduce abreactions.

Bill related to me that when he was working with a Native American group on their alcohol addiction, he found that they would have intense emotional abreactions due to PTSD they seemed to universally experience from childhood abuse. They also frequently fell asleep during sessions due to the lack of good sleep they got at night. His idea was to inhibit delta to try and keep them more alert. What he found, he believed, was they had less abreactions and consequently incorporated the delta inhibit permanently in his method. Bill was devoted to the idea of a painless process. We on the other hand, as previously discussed, found abreaction was an important part of resolving trauma. Our initial response to his findings was a compromise and to initially use delta if individuals displayed the abreactions common to severe PTSD but later turn of the feedback from delta as clients progressed. For us, the ultimate goal was to just utilize alpha and theta as the original research had indicated.

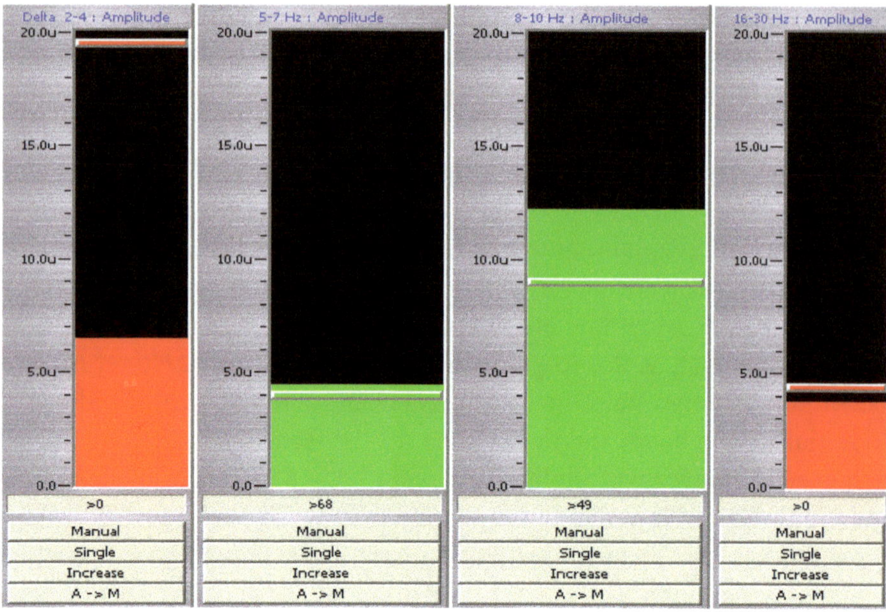

Note in the screen above that this individual has beta below threshold and so we will turn off the beta. The alpha is initially well above threshold and the theta is much lower, with the theta tone only periodically present. The client is listening and feeding back primarily on the alpha tone. In time, as mentioned above, they begin to experience less, and less alpha amplitude and theta begins to increase as shown in the figure below.

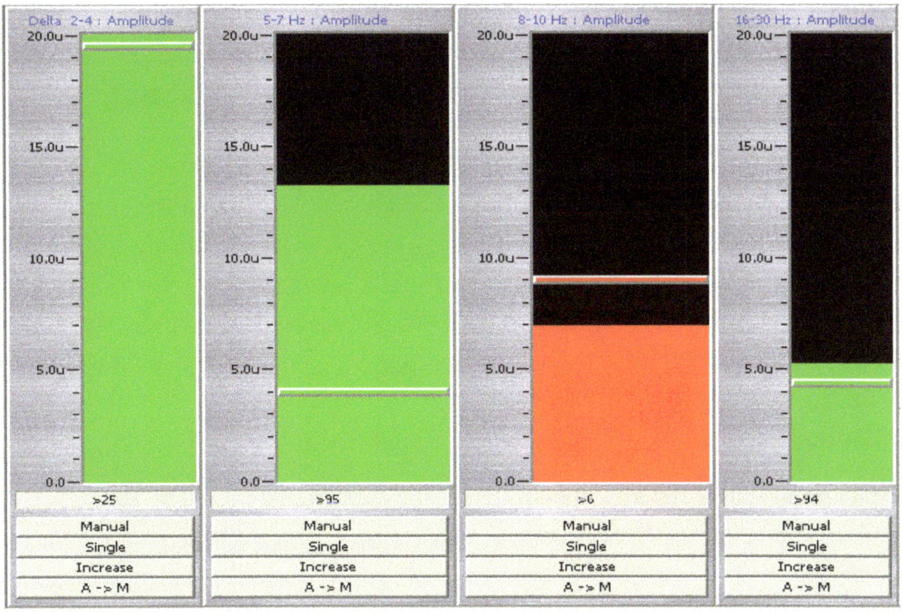

Eventually the theta dominates for a while and delta begins to increase. Soon paradoxical alpha will emerge, and the state of the design screen will flip back to the original pattern indicating a shift in arousal based in fundamental attractor basin patterns in the EEG network fields. A successful session will show many of these shifts.

Setting up this screen correctly is very important. Using auto threshold to set the initial threshold patterns is one of the easiest ways to start. With alpha set at 70%, theta set at 40%, beta and delta set at 10%, the client closes their eyes and their EEG levels are recorded so that the thresholds find the correct levels of reinforcement. After this 2-minute baseline, the session is paused and taken off auto-threshold. The resulting threshold levels will remain the same for the rest of the session. What you have done is individualized their thresholds for the entire session. If you have New Mind software this is built into the system, so it does all of this automatically.

Sensor Placement

The original work by Elmer Green was done at O1 in the occipital region. This location was selected because Elmer's research indicated that this was the location showing the most unique response during the state he called crossover. This was a state that he and Alice Green found to be most conducive to creative insight. Peniston, however, began to experiment with O2 and Oz during his research with veterans with PTSD. Clinicians, based on this research, began using all these sites. Bill Scott eventually moved to Pz, with Peniston's blessing, because the participants kept knocking the sensor off their head when they laid back against the recliner during their sessions. Later the research on the default mode network revealed that this area was right over the self-referencing hub of the brain. At this time, however, the method still seemed to work well so many other clinicians followed suit. As time went by, Nancy White, who was a major early innovator trained by Adam Crane, decided to use P4 because of research published about the relationship of that area to a sense of expanded self-awareness by Dr. Andrew Newberg in 2001. Matthew Kelly (1997), also doing research with alpha theta with Native Americans, did his training at Cz and found he had good results as well. Clinicians continued to innovate in all these locations. So where to best place the sensors? There is no research available to sort this out. There is, however, extensive shared clinical experience.

The prevailing clinical opinion at present is that it is best to train individuals with excess delta or theta in the posterior regions so as not to inadvertently generate too much frontal slow wave during sessions. This is also one of the reasons for doing initial fast wave training before AT training as well. Individuals with too much beta on the other hand may benefit from some more frontal slowing to calm them down faster, so Cz may be a good placement for them. We have found many individuals with excess high amplitude alpha often have difficulty with crossover at Pz but are able to do well at Cz. Of course, these decisions are based on the qEEG brainmaps. It does not hurt to move around to find what is best for each client.

Interpreting The Significance of Training Screen Patterns in Alpha Theta

Dealing With Abnormal EEG Distributions

A considerable number of your clients will come into your clinic with significantly abnormal EEG distributions which make it seemingly impossible to recognize or achieve a crossover on the trend screen. This was confusing to everyone at first, but eventually Bill and I sorted it out. It was Bill who had the brilliant idea of "follow the alpha." Barry Sterman (1996) wrote an article describing what happens to the waking EEG distribution as individuals shift into sleep. Everyone has the same basic pattern, but it becomes distorted by emotional and physical trauma. This is why many people have sleep problems. Key to this pattern is the previously described descent of alpha in amplitude and frequency as we transition into phase one of sleep. The other key indicators are increasing theta and a sharp increase in delta occurring at the same time as alpha reaches its nadir. Unfortunately, many individuals with head trauma have delta that is already elevated from inflammation while those with ADHD have elevated theta, and those with severe PTSD have elevated beta and elevated high beta. The higher amplitude beta is usually high frequency and high amplitude beta in the 20–30 Hz range that has subharmonics spilling over into the 15–20 Hz beta filters. All of this makes it difficult to look for the standard pattern of changes in the trend screen.

To begin with let's look at someone falling asleep in the trend screen below. First, the alpha (the line represented in light blue), rises, and then begins its descent in amplitude. As it reaches a low point around 7 minutes, theta (represented by the dark blue) and delta (black) begin to increase with delta showing a very dramatic increase in amplitude. This is characteristic of a rapid movement into stage 1 sleep. Note that the high frequency beta in yellow, which represents muscle tension, or bracing as we call it in biofeedback, also declines significantly at this point to the normal level of 3–5 μv. Individuals with PTSD tend to show either no decline or an increase in this high frequency (yellow line) activity. They involuntarily brace as they reduce their vigilance level and let their guard down. This bracing is a form of conversion where psychological distress is reflected in the body, and this can take a great many forms. Increasing EMG is one of the most common.

This EEG is overall a low amplitude EEG, as can be seen by the distribution of the first 7 minutes where all the frequencies crowd around the 6–9 μV range but delta amplitude increases to over 15 μV during stage 1. Note the increasing theta as the stage deepens.

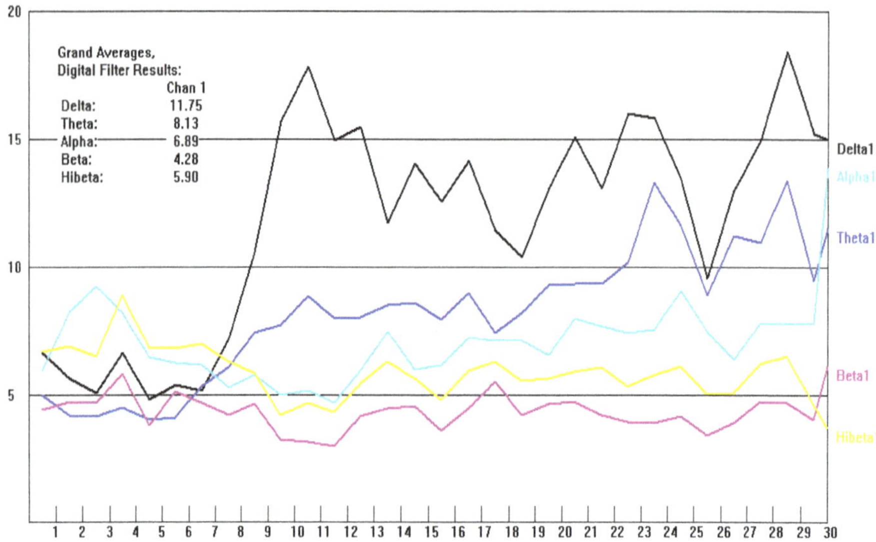

Client with sleep disorder drifting into sleep in first 7 minutes.

Another common pattern is crossover adversity due to emotional trauma, and this is common with PTSD. This individual also has low power EEG, but the alpha begins in normal range around 12 µv. It slowly descends until it touches theta around minute 12 but climbs back up and remains above delta. This individual suffered childhood trauma and was recovering from alcoholic behavior. The reluctance of the client to enter crossover is involuntary. Note high beta starts elevated around 7 or 8 µV and diminishes to 5 µV but begins to increase with alpha amplitude. Alpha represents conscious awareness in this paradigm.

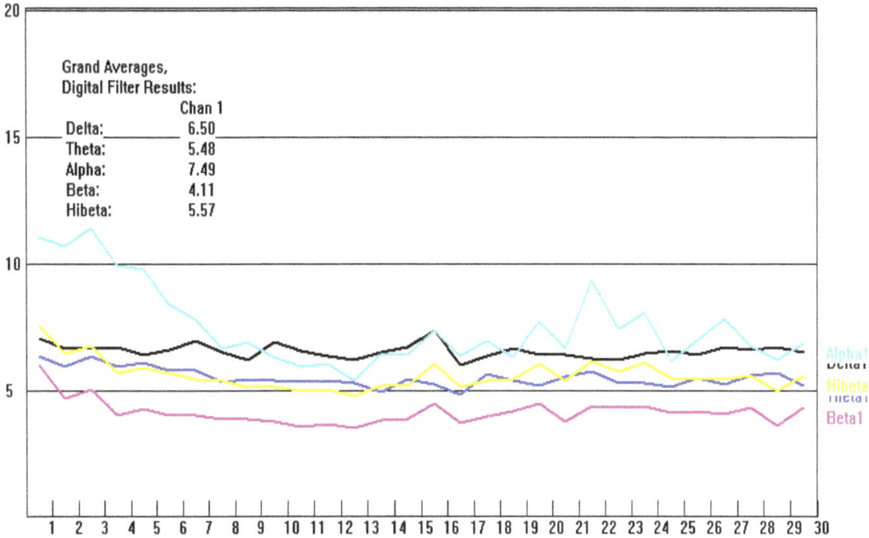

Trauma client adverse to crossover.

This same client experiences a crossover during the seventh session. The seventh session is noted in the original literature as a typical point for first crossover when dealing with significant emotional trauma and after decades of clinical observation we have found it to be common for crossovers to occur within the first two or three sessions for those with minimal trauma. Note the crossover is again around minute 12 and is very brief with an immediate increase in alpha indicating the client shifted in and out of self-awareness for a moment. It was typical for this client to demonstrate an involuntarily exaggerated startle response just as he approached crossover. If you were watching the HRV screen during this process, you would have noticed increasingly poor coherence as the client approached the crossover point. This is related to his resistance to experience and integrate the emotional trauma. The amygdala is becoming hyperactive and signaling danger as cortical and subcortical pathways open allowing memories and emotions to surface for binding and integration. We eventually had to enroll the assistance of an EMDR therapist to get past this point, which is unusual but does on occasion happen with severe cases.

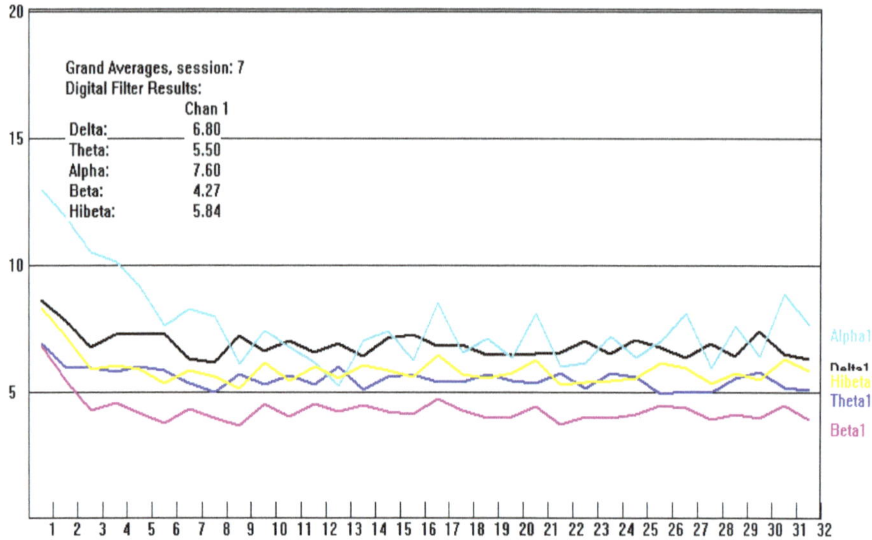

Same client: Crossover occurs seven sessions later.

The next example below shows an individual starting in the normal brainwave amplitude range with delta, theta, and beta, but also with a very high amplitude alpha. This alpha is in the 20 μV range, and this amplitude is at least one standard deviation too high. In the past we speculated that this EEG pattern was frequently due to toxins in the environment, but more recent research shows it is most commonly a response to trauma. This more recent research, involving several investigators and replicated designs, has found that this high amplitude alpha at the 10-20 location of Pz over the DMN to be common with those with excessive "mind chatter" that is often related to complex trauma (Ros et al., 2013; Nicholson et al., 2022). It is not uncommon for clients with this profile to find it difficult to quiet their busy mind enough for the alpha to drop in amplitude. Like elevated high frequency beta this high amplitude alpha is another form of managing trauma. This repetitive idling of the posterior cingulate becomes an effective way of distracting the DMN and preventing input containing uncomfortable emotional input from the subcortical limbic region.

On this client's brainmap we could see that the dominant frequency was not in the normal range, and they had a strong beta asymmetry, a common indicator of anxiety, so we know that this person had metabolic issues relating to anxiety. They also scored very high in the Interactive

Self Inventory in dimensions of anxiety but low on inhibited, impulsive, and perfectionism, all common features of chronic psychological anxiety which, when low, indicate conversion of anxiety into physiological distress. Note the high amplitude of high beta around 9 μV indicating intense bracing as well, which is a further supporting indicator of this conversion phenomenon. The high amplitude beta does come down over the course of the session but so does the theta. As the individual relaxes and muscle tension decreases, theta decreases as well. The individual begins to feel too vulnerable and involuntarily braces in terms of EEG by reducing theta input. The alpha does come down in this session but not low enough to initiate a crossover and unfortunately, we see no significant increases in theta or delta. HRV would be showing increasing decoherence at the alpha low points where the client involuntarily feels unsafe because of emerging feeling related to previous trauma. Again, the low amplitude theta blocks emotional input into the DMN where theta should be higher than average. No crossover experience is accessible in this case.

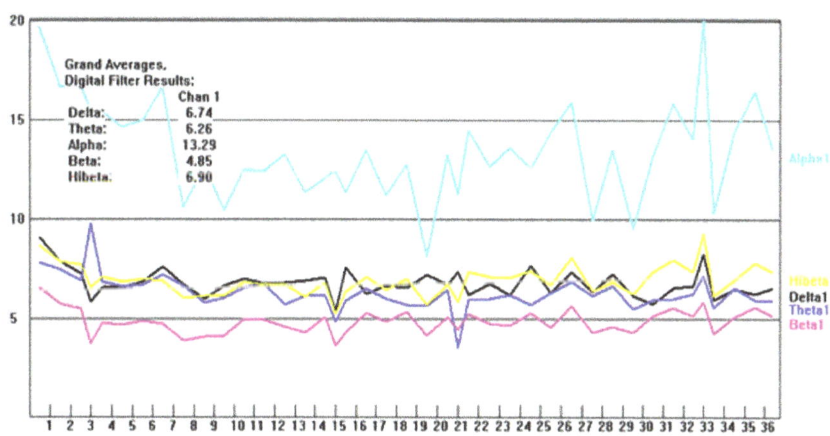

The next case is similar to the last except in this instance we know from the metabolic checklist and lab reports that this individual has low TSH and especially low thyroid T3. This results in a lack of available T3 for enzyme production and consequently a diminished efficiency of sodium pumps (Dietzel et al., 2012; Gilbert & Zoeller, 2010). It also leads to excessive network idling resulting in diffuse high amplitude alpha, but this time it is a very slowed alpha with a slow dominant frequency between 8 and 9 Hz, or even slower. Note the theta is higher in amplitude than the previous example during training, even though it is lower in the map. This low posterior theta is common in anxiety and the lower central delta suggests chronic sleep issues. The increasing theta usually signals stronger emotional self-awareness, and the beta is coming down, but the high amplitude alpha is problematic. Note that the beta follows the high frequency beta as muscle tension reduces. Once again, crossover is still very difficult if not impossible to achieve. Photic stimulation during the session might help but some type of thyroid support may also be important.

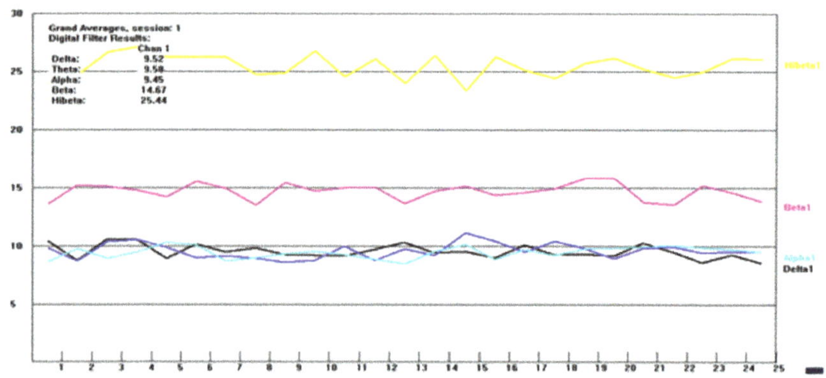

Note Hi Beta (yellow)shows extreme muscle artifact leaking into Beta filters (red).

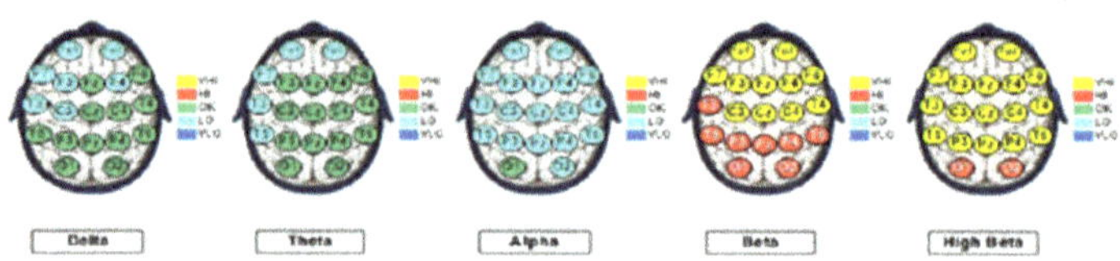

Crossover difficulties due to high amplitude alpha caused by a thyroid condition.

On the next session trend screen we utilized photic stimulation at 7.8 Hz in an effort to generate a crossover after several failed sessions. Photic stimulation involves a strobing light at a predetermined frequency that shines in the clients' eyes. For this we use a pair of glasses that emits the light which penetrates the closed eyelids and stimulates the thalamus at the same frequency as the light. This activity entrains the cortical alpha activity increasing phase and coherence and normalizes symmetry in the theta domain and often the alpha domain as well. The consequence of this increasing theta activity is often, but not always, a reduction in alpha activity. This is likely because it stimulates thalamic gating in a manner to the onset of sleep as reported by Barry Sterman (Sterman & Bowersox, 1981). In this instance, we were able to precipitate a crossover event. The client reported a hypnogogic image without any expectation or coaching, which confirms the liminal state was achieved. We have several other documented cases where this approach was very effective.

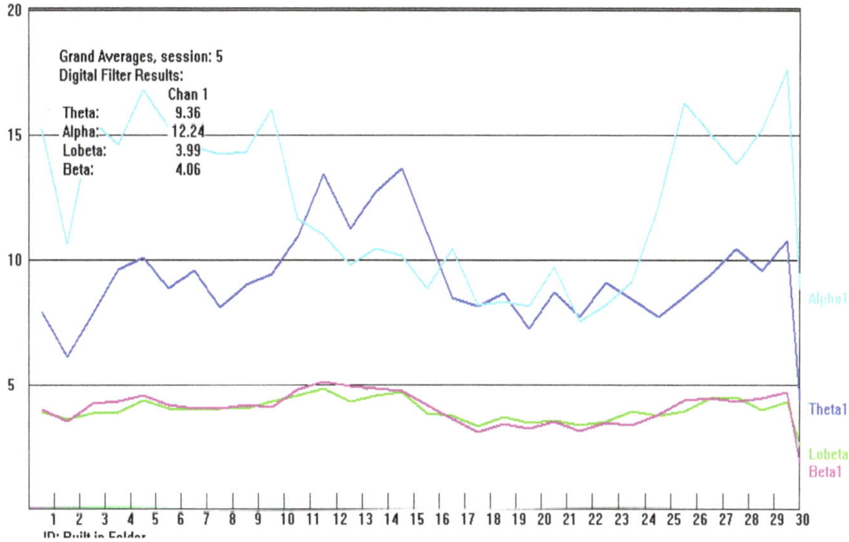

Using 7.8 Hz photic stim to assist with crossover.

The next client is a classic map found in individuals showing many of the features of ADHD. Excess elevated diffuse theta suggests globally diminished cortical blood perfusion. This happens when oxygen is low as well and metabolism is slower. Note that posterior alpha is elevated, indicating significant mind chatter and perseverance activity typical of anxiety, and developing depression. This map also displayed a fair amount of alpha asymmetry confirming the depression. The alpha does not go down during training and neither does the theta increase, resulting in no crossover. This could be either a case of cognitive decline, as in Parkinson's, or an individual who was highly dissociative and heavily involved in daydreaming. Daydreaming is another way of managing trauma, similar to mind chatter, and blocking the emergence of traumatic memories and feelings. Typically, it would be important to begin with training theta down and SMR up at Pz, and practice breathing with HRV to engage the individual with moment-to-moment activities and experience before doing AT. However, if alpha had declined significantly and the theta increased at the same time then we would suspect a crossover event.

ro

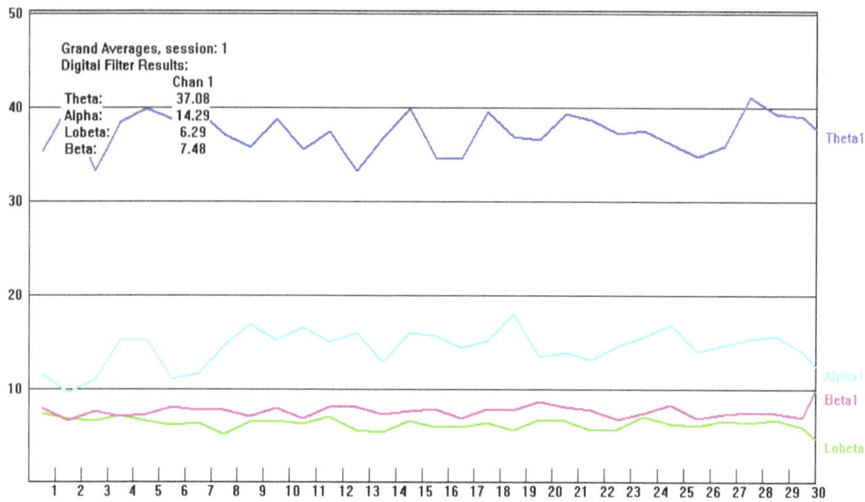

High theta from daydreaming/dissociation presents as if crossover already occurred.

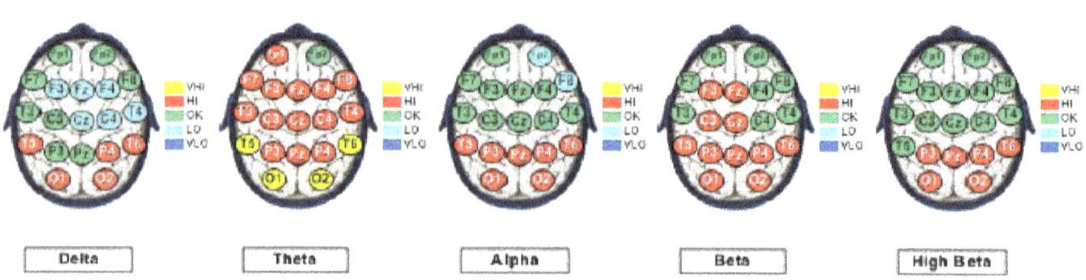

The next case is another good example of this phenomenon of excess slowing and the difficulties it presents for standard AT training patterns. This is a typical profile of an individual with head trauma, and the elevated delta is a proxy indicator of underlying inflammatory process (Knyazev, 2012). There is an elevated amplitude of beta and high beta, and this could only be possible if the individual was clenching their jaw tightly, and in this case we knew from the history that bruxism and jaw misalignment was part of the clients symptoms. Alpha and theta begin at almost the same amplitude. Note that alpha goes very low between 7 and 11 minutes and theta and delta show simultaneous increases. Also note the mild decrease in high beta and muscle bracing. Prior to the decrease in high theta an increase in HRV decoherence would be building and aiding in identifying a potential crosssover. This is likely a crossover moment that could be confirmed by asking the client about hypnogogic images.

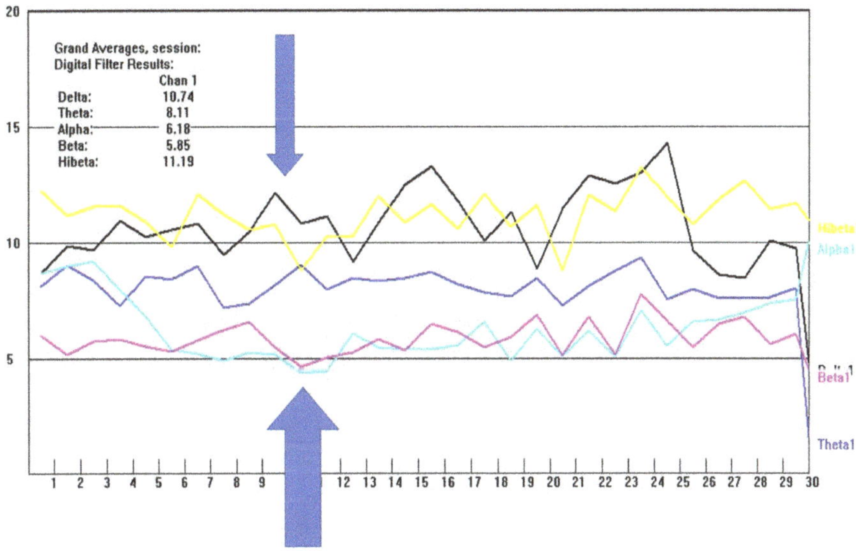

In the next session example below, we can see that we are starting with higher-than-normal delta from a previous TBI compared to the alpha amplitude, and alpha is descending from 10 μV to around 5 μV with a drop in high beta. The drop in high beta indicates a significant reduction in hypervigilance and bracing. The drop in alpha also confirms a reduction in vigilance. Delta is decreasing and theta is increasing only before and after each alpha drop, as if the client is involuntarily avoiding the crossover. There should be increasing and decreasing HRV decoherence or changes in hand temperature if they are being monitored. Around the 29th minute there is a brief touching of alpha and theta that suggests a good crossover may be present in future sessions. No crossover with theta. Finally, around minute 34 we may have a very brief but superficial crossover. So, the elevated delta did not prevent a crossover from occurring.

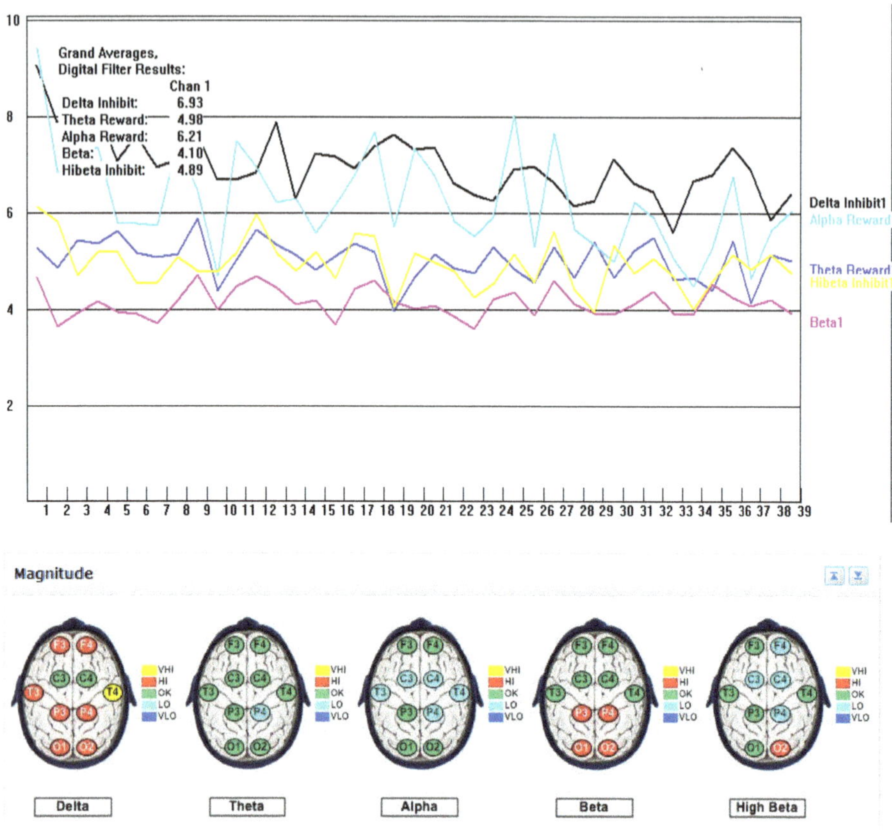

TBI with crossover resistance.

The next example shows high amplitude high frequency beta that could only issue from chronic jaw tension related to TMJ, and no changes in theta and alpha that would indicate a crossover, even though they do crossover due purely to the fact that they are so close in amplitude. These would be false crossovers. Note also that the very high amplitude beta is an artifact of subharmonic leakage of high beta frequencies into the beta frequency filters due to signal processing artifact. This beta is not real. An individual with this profile would benefit significantly from posterior alpha uptraining and extensive HRV training as well as practice in breathing techniques to reduce chronic hyperarousal. It would also be helpful to do some EMG biofeedback if possible, to regain conscious control of the excessive muscle tension.

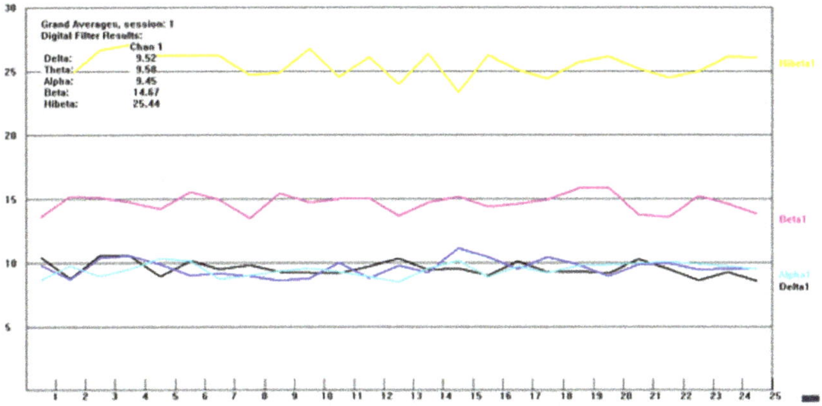

Note Hi Beta (yellow)shows extreme muscle artifact leaking into Beta filters (red).

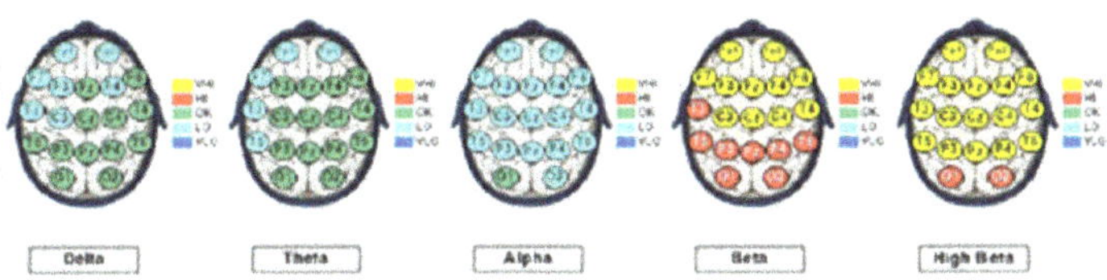

Finally, in this last trend screen below we see an optimal session with multiple crossovers occurring after this client had 30 or more AT sessions. The HRV should be indicating that the client is maintaining a good level of HRV coherence and/or normal hand temperatures or even improving coherence as the session progresses. The constant weaving back and forth of the alpha and theta indicate the client is well established in an ongoing liminal state providing optimal integration opportunities.

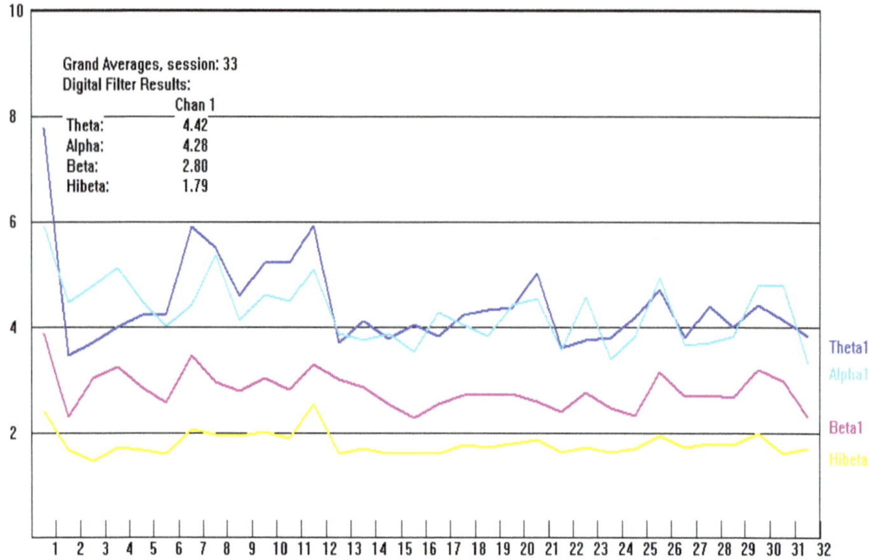

Optimal consistent crossovers indicating continuous twilight state.

Chapter 9

Confounds To NFB Training

When individuals come to a clinic for NFB they typically carry a great number of assumptions and erroneous ideas with them that are an impediment to NFB. One of the most difficult to deal with is the "Magic Bullet" theory of medicine (Whitaker, 2010). This idea has been documented to be a major concept in medicine that has led us down a very dark road. The idea is that the doctor or healthcare professional carries an arsenal of magic bullets that provide an immediate solution to your suffering. Each form of disease has a specific solution that can be resolved with a particular medicinal prescription. If any solution does not work right away, then it probably doesn't work at all. I call this the John Wayne school of medicine. "Just patch me up Doc, and I'll get back on my horse and get my job done!" Medicine has never worked this way, except for antibiotics, and healing never has either. It often takes a considerable amount of time to recover from mental or physical trauma. In a society obsessed with production, consumption, and immediate gratification, we do not provide room and time for healing and especially non-physical healing. An especially good example is grief. The work of Elizabeth Kubler-Ross revealed in the last century that as a culture we do not even acknowledge grief and the stages it requires. It has almost become invisible to us as a culture and yet at the same time it continues to undermine the health and performance of millions of people for years on a regular basis. People suffering from grief are frequently misdiagnosed as depressed and then they are medicated. Ironically, the unfortunate consequence is that they often never resolve their grief because the medication blunts their affect and prevents them from experiencing the grief and working through it.

Clients come hoping for immediate relief and if relief is not immediate their confidence is often shaken. These clients have often exhausted a considerable amount of their financial resources as well as their hope and their patience before they arrive. The neurofeedback clinician's office is frequently a stop of last resort in a journey of pain and suffering that many other professionals fail to acknowledge or effectively address. Relief from chronic disease and complex trauma is rarely found in mainstream medicine or counseling that is covered by insurance.

Now, research like the ACE study is telling us how these two forms of affliction co-exist and feed upon each other. Sadly, most professionals are 17 or more years behind on research findings. This relationship is still not well-understood in the clinical world. There is a powerful bi-directional relationship between the mind and the body. This is why any practitioner should be taking a bio-psych-social perspective on their client or patient. When I look at the literature on this topic and how much of it is available, I am always surprised by how little so many professionals understand what is right in front of them. In addition to the bio-psycho-social perspective, professionals should also be adopting the perspective of functional medicine that is emerging and consequently we promote the concept of functional neurofeedback. This means seeking the source of presenting symptoms and not just treating symptoms themselves.

Functional neurofeedback is an integrative model of care that offers a client/patient focused approach to addressing disorders of the brain and central nervous system. The goal is to provide personalized client/patient-centered care. We recognize that each symptom or differential diagnosis may be one of many contributing to an individual's illness.

Functional neurofeedback clinicians complete a comprehensive intake and assessment and take a comprehensive health history. The clinician uses this information to better pinpoint the root cause(s) of brain, central nervous system, and mental health disorders that include but are not limited to diet and nutrition, physical health (metabolic and physiological), stress levels, environment factors, lifestyle, and sleep hygiene.

The goal of functional neurofeedback is provide QEEG brain mapping for assessment purposes and brainwave training (neurofeedback) to address disorders of the brain and central nervous system in order to promote a healthy lifestyle.

There are those in the field of neurofeedback who are burdened with this older magic bullet paradigm as well. This can be a source of lack of progress from the start of the first day of training. These clinicians with this more archaic perspective are constantly in search of one protocol for one symptom solution. "What is the protocol for ADD?" they ask. As we have discussed, not only is the concept of ADD deeply flawed, but the idea that there is one protocol for it is also just as deeply flawed. It is only natural for a clinician to be driven to discover "the solution" to someone's problem. Expecting a simple source of the problem or a simple solution, however, is usually unrealistic. The human brain and body are some of the most complex structures in the known universe and when things go wrong, it is usually very complicated. There are usually many factors contributing in concert, and they often involve multiple lifestyle factors such as diet, sleep, and exercise, as well as quality of personal relationships. Using multiple assessments along with a qEEG and detailed interview are critical to understanding the client in depth and these assessments need to be strategically repeated to track client progress. This collection of assessments allows the clinician to review the brainmap with greater depth of understanding regarding the client's condition and the factors that contributed to that condition. I have frequently seen clinicians disregard the assessments we provide in order to quickly get a qEEG brainmap, thinking that the map is all they need to understand their clients and come up with the exact perfect protocol. Nothing could be farther from the truth. So, some of the first impediments or confounds to good NFB outcomes are the attitudes and beliefs that the clinicians and their clients bring to the table.

NFB is neither invasive nor coercive. It is a form of involuntary learning (Birbaumer et al., 2013) called operant conditioning. This is why we tend to use videos for training rather than have clients willfully straining to change an image on a screen, as in the early days of NFB. Willful struggling and straining only makes things worse. Things go better when we work with the biological system and do not try to force it. One of the important things that our data has repeatedly shown us is that healing does in fact usually take time. On average, it takes 15 training sessions to generate significant documentable changes in the trend screens that result in significant documentable changes in symptoms. We have tens of thousands of symptom trackers and pre/post maps to prove this point. We have documented it with all forms of NFB that we are aware of currently. This does not mean that things can't happen faster. We have many trackers that show very large changes in symptoms within five sessions. But those changes will not be sustained unless they are consolidated by further sessions. This is why we urge practitioners to be wary of claims of the magic bullet and quick fixes. This approach has already failed in both medicine and in NFB. NFB always works if allowed because learning always works if allowed. This is why we have school lunches and school counselors. Children cannot learn in a class if certain conditions are not met. If they are too tired, hungry, cold, frightened, or intimidated, they learn poorly if at all. What slows progress down in NFB is externalities such as abusive jobs or family environments or poor lifestyle habits. Many clients secretly hope they can continue to maintain their old unhealthy lifestyle and get "fixed" by the NFB. This is simply not possible. Changes in lifestyle will always be important when addressing suffering and they will be part of the solution. Most medical doctors are very aware of this and are usually very frustrated by expected poor compliance. Our biggest impediment to healing is ourselves and our automatic habits of thinking, feeling, and doing which we acquire over a lifetime, and which lead us to ill health. This is the at the heart of the ACE study findings.

It should be emphasized to clients at the outset that they are responsible for their own healing and not the clinician. The clinician is the technician who will provide them with the resources to recover. The client must do the work, must do the learning. They must be patient with the outcomes. This is the bio-psycho-social approach to healing. The clinician must help them identify the lifestyle patterns that are contributing to their suffering and assist them in learning to overcome them. This is when NFB works best, as an adjunct to healing changes in lifestyle patterns and behaviors associated with suffering and ill health. This is a challenge because it requires empirical measures and tracking of these multidimensional factors across time. We must measure and quantify metabolic symptoms, psychological symptoms, cognitive problems, and physical health problems and look for patterns of lifestyle and behavior that are contributing to them. We must recommend changes and support them through training, practice, and monitoring techniques. When done properly in conjunction with NFB, profoundly positive changes in symptoms usually follow. When NFB is not working well it is usually because of a problem that is metabolic or one that is emerging from social relationships in the family or at work.

Drugs

Drugs can aid clients initially in dealing with acute phases of suffering, but they often become an impediment when applied to chronic conditions. They can contribute to sustained symptoms of chronic conditions. When doing NFB it is usually necessary to taper slowly off drugs and get them to minimal levels or remove them completely. After decades of clinical observation, we have found at New Mind Center that clients who reduce their dosage levels too rapidly experience severe side effects. Unfortunately, many prescribing doctors appear to want to rush the process, and this almost always ends in disaster. We recommend and request from prescribers, small reductions in dosage with each reduction at least 3 weeks apart. Rapid reductions in dosage at the same time NFB is in progress can create a synergy that can push clients too fast and too hard.

Clients will begin to experience many of the listed side effects of a drug as they continue to train because their primary symptoms abate, and their biological systems improve so that their usual dosage becomes too high. If clients do not taper off, they may not get better and may even begin to feel worse. At this point they will blame the NFB. This should be avoided at all costs. Inform clients of this problem at the outset and be sure they have a list of side effects posted somewhere in their house. Get written agreement to work with you in this area. Be sure they have access to a medical practitioner who can help them manage and reduce dosage. Our experience is that the Physician's PDR is often unreliable for this process. As mentioned, often we see physicians reducing dosages too rapidly and we can clearly document the effects in our system. This phenomenon has been widely published and acknowledged in the psychiatric literature (Whitaker, 2010). Clients need to know this as well. Those who start out without drugs, on the other hand, frequently do much better and recover more quickly from their symptoms and their suffering.

As I have mentioned, it has been noted in the literature that drugs tend to blunt affect. This reduces client access to emotion and this in turn reduces their ability to think clearly and make good decisions. At the same time, too much distress can result in the same problem. A balance must be struck between these two issues. Emotional awareness is crucial for integration from psychological trauma and recovery (van der Kolk, 2014). This is also why we start out immediately teaching individuals with heart rate variability (HRV) and other biofeedback techniques how to calm their CNS. It allows them to manage the anxiety underlying their distress, provides them with a sense of control and leads to faster integration.

The drugs that are the greatest impediment to NFB progress are opioids and benzodiazepines and these are also two of the most abused drug categories. They are addictive and this results in clients underreporting their use. SSRIs on the other hand have a milder affect except when combined with a large number of other drugs. Large combinations of drugs with high doses are a warning sign that the client is already overmedicated and poorly managed. They are also poor candidates for NFB unless they are highly motivated. Those clients on more than three psychoactive drugs tend to train poorly and make slow progress.

Sleep

Insufficient sleep can cause a host of symptoms, yet sleep has been ignored in the past as a significant contributing factor to poor health and mental disorder (Walker, 2017). Fortunately, in the past two decades it has become more recognized as one of the most important activities we engage in. Those with insufficient sleep usually have unusually elevated eyes open delta. For those with long term chronic sleep issues, very low delta is often present on the map, especially in the eyes open condition. There are over 80 parasomnias, and a major portion of the public gets poor quality sleep or insufficient sleep. NFB is one of the most effective methods for restoring quality sleep, and AT training in particular is highly effective.

Working the late shift is also extremely detrimental to both physical and mental health. We frequently must arrange for medical personnel to change their shifts before we can do effective training. Backsliding is a common feature. They may do a couple of good sessions and then retreat on the next one to their prior trend screen pattern. If they don't change shifts, they also fall asleep too easily at the beginning of their training session, even when doing eyes open training. Their delta rarely diminishes over the session and often increases.

No amount of training or any clinical procedure can overcome insufficient sleep. We recommend that clients get Fitbit or Oura Ring and begin tracking their sleep. We can then view their sleep data to monitor their progress, changes, or backsliding. This helps them to get more actively involved with tracking and improving their sleep as well as following sleep hygiene guidelines.

Eating Habits and Nutrition

It has been well documented that the modern prepared foods that the majority of Americans eat are highly contaminated with untested chemicals and toxins (Hyman, 2020), including pesticides. The adipose tissue of most individuals from birth is highly contaminated with these toxins. Manufacturers engage in deceptive practices to hide the contaminants in their products. When children are highly physical and active, they burn fat, and this releases toxins into the bloodstream. A sub-concussive event can result in a brief breach in the blood brain barrier that can allow these toxins access to the brain. This can initiate a cascade of inflammatory processes that become chronic and with a deleterious effect on neural function as well as neuroendocrine disorder or NED. We review the literature on this in the section on metabolics and the oxidative stress and qEEG research.

Even without concussions or sub-concussive events these contaminants inflame the intestinal tract and result in excessive membrane permeability. This is usually mirrored in the brain membranes as well (Perlmutter, 2015). This permeability leads to an inflammatory response that often ends in an autoimmune cascade (Paray et al., 2020; Wang et al., 2019). This in turn affects endocrine support of neuronal function and results in a degradation of the neuronal system. Research has documented that this inflammation leads to anxiety and depression (O'Mahoney, 2015). The resulting inflammation can be tracked with the qEEG as we demonstrate in this book. This condition is often what clients are suffering from when they come to your clinic as documented in the metabolic checklist and the brainmap. To counteract this process, clients must learn to change their diets and make better food choices. It is essential they eat only organic foods as much as possible and meals prepared from these sources. This requires an open mind and experimentation. For many clients, this is as difficult and emotionally challenging as changing their religion. They must learn to eat more vegetables and fruit and less meat. There are many good books on this that clinicians can provide clients for guidance. Sometimes even cooking lessons are required to help them make the transition.

Supplements are also usually important to recovery and the NFB process. Most clients come in with serious nutritional deficits. Prolonged or toxic stress leads to chronically elevated cortisol levels, adrenal fatigue, elevated homocysteine levels and immune system suppression. This in turn leads to deficits in magnesium and zinc. It is well known that the majority of individuals have a chronic deficit of vitamin D, B12, and omega 3 fatty acids as well as severe magnesium and zinc deficits. These mineral deficits affect neurotransmitter production and suppress critical enzyme activity. Without these resources, it is more difficult for the brain to alter structure and function in response to NFB training. The New Mind map system provides recommendations of nutritional supplements based on the qEEG profile. Most nutritionists who have reviewed the output have told us that it is usually similar to their own recommendations after assessment. It is meant to be an aid to them as well as other healthcare practitioners in determining nutritional recommendations. It should not be blindly followed. In addition, the metabolic checklist is designed to be used with the Apex nutritional system and is designed to interface with their products. Practitioners can find a one-to-one correspondence between the metabolic checklist and their assessment system and its categories.

Exercise

For over a century, scientists have been conducting research on exercise and its effect on human health and functioning. The message has been consistently clear and is becoming louder every decade. Lack of exercise contributes significantly to poor sleep, poor eating habits, poor microbiome profile, poor neural functioning, and low energy levels. Like sleep there are now vast amounts of studies to confirm these issues. Nevertheless, most people do not get adequate exercise, and this contributes significantly to all of their physical and mental health problems. That said, it is notoriously difficult to get clients to develop an exercise program and stick to it. For those finding it difficult to discipline themselves in this area, exercise programs like Jazzercise, Pilates, Tai Chi or yoga classes can be effective because of their social support. Finally, if that is overwhelming initially, a Fit Bit biomonitoring device can help to effectively change a client's entire attitude toward walking and other forms of exercise. These devices are very good at engaging people to compete with themselves and discover that extra steps and physical tasks can be a positive experience rather than something to be avoided.

Relationships and The Family System

Unless they are educated and trained in family systems, it appears most health practitioners are only dimly aware of their importance. There is presently a vast literature on this topic in several disciplines going back to the 1940s (Soutar, 1996). The ability to develop relationships is forged through early socialization in the family system. This is a biological process as much as a socio-emotional and psychological process. The family is not only a fundamental economic unit but also a fundamental socialization context. It transcends all others in its impact on the formation of personality and behavior. Judith Lubar was the first to document its effect on NFB (Lubar & Lubar, 1999). It can have a profound effect on training and even completely mitigate it. We have written about this in the book Doing Neurofeedback (Soutar & Longo, 2011; 2022) and it is part of the BCIA Blueprint of Knowledge and consequently the New Mind Academy Web course. I developed this theme in detail in the chapter on the psychology of NFB. Individuals with medical problems such as severe migraine or Alzheimer's, addictions, psychological disorders, all profoundly impact the family system. Financial problems, death, divorce, relocation, and a host of other social circumstances can also adversely impact the family system. The effects of these events impact the formation of personality and the social self as well as the ability of individuals to interact in a manner that is socially effective and allows them access to resources they require to thrive.

Individuals living in these relationships and family systems grow and form neural networks and systems that reflect their environments. We not only become anxious, but we also grow an anxious brain, and it operates differently than a healthy brain. This cannot be changed overnight. The ISI measures the dimensions of behavior that emerge from social interaction and can indicate if client symptoms issue from an adverse socialization process involving toxic stress or not. Once identified, a clinician can work to help clients reframe their beliefs and change their habits of thinking, feeling, and doing. NFB provides the plasticity and enhanced self-awareness that is required to emit these novel behaviors as a consequence of natural insight and creativity of response or action. Unfortunately, the family system pushes back and challenges any change. New behavior in the family system causes tensions as it challenges the natural attractor basin or tendencies the system has evolved. This in turn shows up on the trend screens as a poor training session.

Below is an example of a boy who was suffering great difficulties at school because of an emotionally abusive teacher and the change that resulted from his removal from that environment. Note the highly elevated theta that is minimally responsive to training and the post trend screen a few weeks after changing schools. Judith Lubar (Lubar & Lubar, 1999) documented effects such as these with family events such as divorce, alcoholism, and death in the family. If such events from the family system can entirely short circuit NFB training, then it should clearly be of paramount importance and in the forefront of every clinician's mind with each client at each session. When trend screens change, clinicians should ask questions with direct probes. Do you like your teachers? Did something bad happen today at work? Did you have a fight with your spouse?

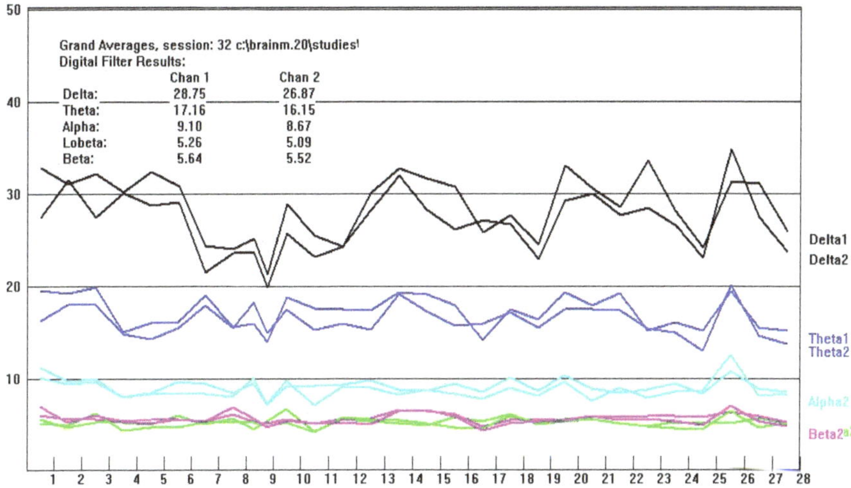

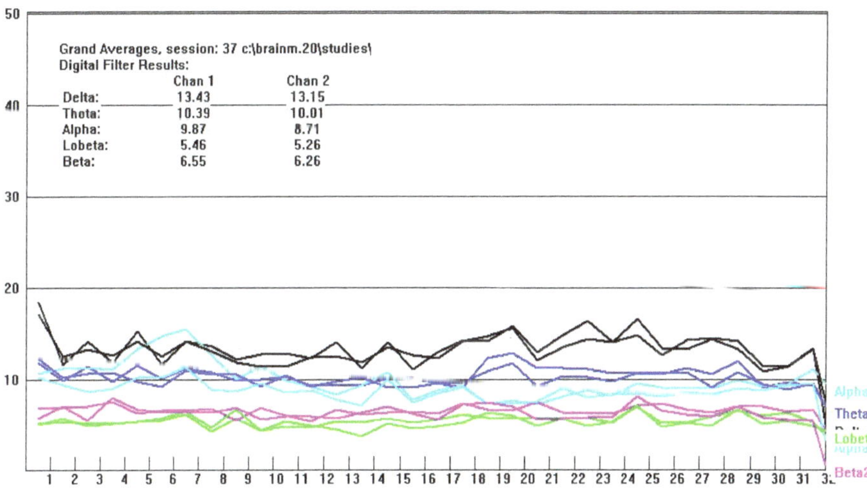

Top session summary chart shows high alpha and theta in response to emotional distress and the bottom chart shows integration after 37 sessions.

The Occupational Hazard and Dysfunctional Job Environments

The workplace is very similar to the family system and individuals tend to adapt roles in these environments that they had or saw modeled for them in their family of origin. The worst of this is usually a consequence of unresolved socialization trauma. Fellow workers and bosses can be devastatingly abusive and subtle about it. Clients from similar families can fall into roles of the victim without even realizing what is happening to them. Clients may be overwhelmed with the abuse they suffer but unable to deal with it because the boss or the owner is part of the problem. The option of leaving the workplace may not seem available due to financial circumstances. It is important to uncover this situation as early in the NFB cycle as possible. Often alpha theta training can open clients up to insights that can lead to unexpected possibilities for change. It is not that uncommon for clients to seek new employment once they understand the effect it is having on their mind and body, especially when they see it showing up in a concrete form on their trend screens.

More on Metabolic Analysis and Confounds

The NewMind metabolic checklist was developed in recognition at the clinical level that certain metabolic deficits were interfering with NFB training and outcomes. We became particularly aware of this problem training women with hypothyroid problems. They consistently demonstrated significantly elevated global alpha with frequency slowing that placed it in the range of 8 Hz in the 1 Hertz bins analysis maps. This alpha was stubbornly unresponsive to NFB using single channel protocols, two channel protocols, Z-score protocols, IFS protocols and any other intervention, such as photic stimulation, that we also tried. Once we saw the pattern, we sent clients to functional medicine specialists for treatment as the acute care physicians failed to recognize or adequately address this problem. When clients were properly supported with supplements, proper diets, and when necessary, thyroid hormone supplementation, these clients responded very well to training. We wondered how many other metabolic issues were interfering in the difficult cases. I brought this up on a major qEEG listserve and was authoritatively informed, to my surprise, by one of the key people in the field that no such problem existed. Upon reviewing the literature, I found large numbers of publications specifically identifying thyroid issues as something reflected in the EEG. So much for the concept of "authority" and "factoids" in science. For this reason, I recommend that everyone challenge and check all authoritatively given "facts" in this field.

As a consequence of this outcome, we developed a metabolic checklist for assessment purposes in order to identify any metabolic issues that may be affecting the qEEG analysis and interfere with NFB protocols. We did this by identifying the major metabolic categories that typically impacted the qEEG, and pulling from the medical literature the symptoms typically associated with these metabolic categories. We then listed the symptoms and using Likert like scales asked for responses regarding frequency and severity of symptomology experienced from each client. Once we had 3000 checklists and corresponding qEEGs from over 500 clinics across the country, we statistically analyzed the data. We reviewed the distribution of the disorders and normed the data, so we had a clinically normed symptom checklist. Significant scores in any one metabolic area encourage clinicians to look for their impact upon the qEEG. Patterns are beginning to emerge, and they are supported in the literature. Below is the distribution of metabolic problems in a typical NFB practice.

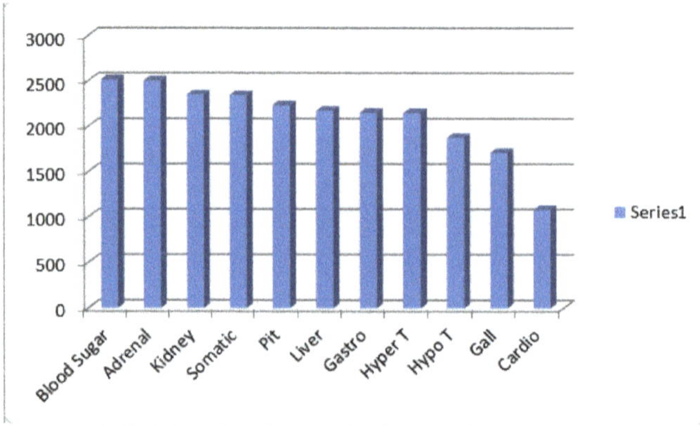

	N	Mean	SD	%
Blood Sugar	2515	16	15.8	88%
Adrenal	2502	18	14.6	87
Kidney	2348	13	10	82
Somatic	2342	18	16	82
Pit	2232	14	6.7	78
Liver	2173	16	14	76
Gastro	2152	17	14.7	76
Hyper T	2147	12	9.5	75
Hypo T	1876	12	9	65
Gall	1710	12	10.6	60
Cardio	1082	13	10	37

Total *N* =
2848

As you can see, a large number of the clinical population have a significant score. Theoretically, a healthy person should have no score or minimal score, so those with significant scores are very compromised physiologically. We have found that the higher the score, the less responsive to NFB they are and the longer it takes to train down their symptoms. This clearly argues against the idea that it's always the protocol or neurofeedback that is not working. If the body and the brain do not have the resources to respond, then there will be minimal response.

I believe we have identified a pattern of general physiological response to stress in this data. Initially stress affects adrenals and pancreas showing up in testing as elevated cortisol levels and glucose instability. Over time the microbiome is affected by an elevated acidic profile and intestinal T joints begin increasing membrane permeability as zonulin levels increase along with intestinal inflammation (Fasano, 2012; Perlmutter, 2015). This occurs, according to the literature, in the brain blood barrier as well. As the immune system responds to foreign proteins in the interstitial fluid and blood stream, an autoimmune response evolves which burdens the thyroid. We then see hyper and hypo thyroid responses. This reduces available thyroid T3 which in turn has a profound impact on endocrine and brain function. Cardiovascular problems emerge due to global inflammation and failure of thyroid regulation on blood pressure. By this time excessive global muscle bracing in response to stress has shortened and degraded muscle tone and health as well. Although summarized here in somewhat oversimplistic terms for convenience, these patterns can be matched to shifts in the qEEG as it gradually transitions through progressively degrading oxidative stress stages that reflect Selye's alarm response and exhaustion stages of stress model and matches our stress model presented in detail in the chapter on physiology.

As a consequence of these findings, we refer clients to functional medicine specialists when their global metabolic score reaches a high level in the order of 150 (approximately 2 SD) or higher. We continue to train but on a reduced schedule to maintain treatment continuity and oversee the client's progress with the progress tracking system as well as pre-post maps and the New Mind Computerized Performance Testing (CPT) of cognitive functions.

Chapter 10

Assisting The Post Trauma Transpersonal and Spiritual Transformations

The Journey From Trauma Integration to Transcendence

To begin to extend the theoretical infrastructure of the therapeutic process beyond the standard domain of the traditional, purely psychological and perhaps sociological realm, it is important to begin with fundamental areas of research and therapeutic implementation that overlap both the spiritual and the psychological domains. To some degree existential psychology as presented by Viktor Frankl has already begun that journey. Fundamental to the existential approach is the dilemma of suffering and death, and making sense of the trauma it delivers to the experience of every human being at some point in their life. The qualitative experience of grief and its transformative effects on the psychology of every individual who encounters it, regardless of their capacity for self-centeredness, usually demands a consideration of "Why is this happening to me?" Experiences leading to grief typically suggest to the individual that they too are vulnerable and mortal. A momentary reflection on one's mortality, however brief, has a powerful tendency to force the individual to peer into the potential infinite darkness of personal extinction. For the majority, if the census bureau is accurate, this also includes reflection on their religious or spiritual reference points. Freud would argue, and many agree, that death (Thanatos) is always in the background acting as a powerful reference and informing the most fundamental aspects of personal identity. In this sense, like sensory pleasure, it is a primary motivator for much of behavior, albeit in a disguised form. Making mistakes or losing in most instances begins to put performance pressure in any context, whether it is a game, a contest, a social performance, or a physical struggle for survival. The challenge to the ego always carries with it a potential loss of some precious identity feature or even full annihilation. This challenge is fundamental to the concept stress itself at both the biological and psychological level. Regret and grief play critical roles in this process, and they are key features of the impetus to explore the meaning of life and reality itself. Grief and loss in particular are extremely well researched and key features of trauma. So, it would seem that grief would be a powerful springboard into theoretical considerations of where to transition the clinical process into more spiritual processing.

Grief: A Key Dimension

Most trauma resolution involves loss and grief. ISTSS acknowledges this but defines grief as an independent dimension that may or may not be a part of complex trauma (Forbes et al., 2020). Klurfield et al. (2020) found grief was statistically independent and distinct from PTSD and consequently one dimension of it. But they also found that grief applied to role losses especially in divorce or job loss. Papa et al. (2014) found that grief is not a unique response to loss of a loved one but is a common phenomenon across many types of loss. This is in line with Kubler Ross's later thinking on the topic. Neria and Litz (2004) found grief is a distinct individual, social, and relational experience. There has been insufficient examination of this issue. For many, grief from a loss is short lived. Recall, regret is related to loss and a key motivation in decision-making theory (Kahneman, 2011). Complicated and acute grief with major depression is almost indistinguishable from PTSD, and how one is related to the other is still unanswered and poorly understood in the literature (Zisook & Shear, 2009). I would propose, based on decades of clinical experience, assessments, and EEG data that they are two sides of the same coin.

Most people I have worked with who have complex PTSD (CPTSD) are grieving the loss of their safety and security, their identity, their self-respect, and their self-esteem. Their locus of control (Bandura, 1997) is externalized, and they experience themselves as powerless victims lacking in self-efficacy. They are grieving the loss of a world that everyone around them appears to possess that contains all these features of safety, love, security, and self-esteem they have been denied. While PTSD calls into question the reliability and safety of an individual's world, most people recover.

CPTSD on the other hand usually involves a repeated erosion of an individual's fundamental confidence in the world and in themselves at the existential level. There is a fundamental sense of existential betrayal. If they really deserved love, they would have not been visited with their dark moment of destiny. For many, this is a loss of belief in the world, in themselves, and in any God or higher power they may have initially believed in, and which deserved their faith. They are angry at the world and the universe or God for visiting on them the experiences they have had, and that calls into question their own value because they feel alone in that anger. Shame, worthlessness, and guilt enter the picture as well. There are feelings and sensations outside their control that have begun to undermine their basic confidence that they can manage things as they are. They fear it reflects a fundamental defect which they carry. This involves an existential doubt which derives from a scientifically unmeasurable dimension of qualia.

These individuals feel consigned to an eternal destiny where they can never really allow themselves to love or trust others in the world again, and perhaps even the world itself. So, it is grief that is at the heart of trauma. There is a loss and bereavement process at hand. Those who were loved and supported from early in life through the various challenges of life often can transcend the loss and move on. This is the finding of the ACE study. Those who failed to have that support and developed without it are more likely to succumb progressively to growing challenges or a sudden event/moment of horrific experience. The resistance to existential trust is amplified and snowballs through their daily experience until it becomes like the story of Sisyphus.

Grief in the form of bereavement is a biological phenomenon expressed by mammals and consequentially an existential event predating our species. Freud saw Thanatos or fear of death as fundamental to our species, yet at base it is an expression of fear that all mammals also carry. Animals can recover from loss, but for Freud, humans internalize it in the unconscious in a manner that cannot be easily resolved (partly due to our social DMN). It can become personalized into a form of defect. It has to be externalized socially to be resolved through social expression to someone who is really listening and—perhaps through mirror neuron action—forgiven. In this case, forgiveness may only occur through the expression of the faces and body language of those around us.

Once again, the problem for a science dominated by the materialist agenda when dealing with this phenomenon is that feelings exist but are qualia and cannot be quantitatively measured like objects, so they must be presumed to not exist or at least should be disregarded in any model. These qualitative phenomena of feelings include shock, disbelief, and denial of an occurred event, sadness, despair and loneliness, anger resentment, regret, guilt and shame, helplessness, insecurity, fear related to anxiety, numbness, emptiness related to depression and so on. Grief experience, like spiritual experience, is hard to measure.

Nevertheless, grief does have directly measurable physical symptoms of heart palpitations, headache, weight loss at the most immediate basic level, and more indirectly measurable symptoms of insomnia, lightheadedness, fatigue, nausea, difficulty concentrating, episodes of crying, aches, and pains. These symptoms are, however, shared with many other categories of disorder involving distress of the CNS as well as with other diseases. In fact, the anxiety that is common to grief has dozens of associated physical symptoms. So, grief remains a fundamental experience of life like happiness that belongs to the realm of qualia.

The research on grief shows that for the most part it is something that comes to be managed and never fully resolved (Zisook & Shear, 2009). Complicated and acute grief with major depression is almost indistinguishable from PTSD, and how one is related to the other is still not answered or understood in the literature. Psychotropic medications have little impact on complicated grief. It often involves suicidal ideation. The major depression of grief is undistinguishable from any major depression, and consequently is often misdiagnosed as depression. Bereavement makes depression more intense and complex in many cases. By 2009, there were no major psychotherapy studies focusing specifically on bereavement related depression.

Few authors and researchers have articulated the role of grief in the human experience as well as the renowned psychiatrist Elizabeth Kubler-Ross. In the mature development stage of her work regarding death, Kubler-Ross (Kubler-Ross et al., 2005) extended her stages of death and dying to all forms of loss and grief. Death of a person or a stage of life or of an expectation all carry a finality that exposes the individual to the grief process. The traditional answer was acceptance through religious belief and ritual. Religious and spiritual traditions are the most primitive and fundamental socio-cultural and psychological efforts to deal with existential fear, loss, and grief. Their roots extend far back in time beyond even oral tradition. They reflect an innate human ability and method for dealing with trauma, grief, and loss.

These traditions suggest we have this innate capacity to overcome trauma and loss. They contain both cognitive-emotional components as well as rituals that induce altered states of mind. They utilize dancing, singing, hallucinogens, isolation, self-mortification, and a host of other methods to induce an altered state of mind where insight, integration and transformation can occur. They include dreaming, lucid dreaming, and dream therapy. These same states have the dual effect of integrating trauma and resolving grief by reconciling the individual to their existential plight, by reconnecting them with their ground state of being at every level of experience. Experience is a key component if not the definition of consciousness (Koch, 2019). AT training is deeply connected with these methods because it arises from the same traditional sources but evolved through contemporary technology rather than primitive technology. This is clearly expressed and evident in the writings and work of Elmer Green. It was his intent to adapt technology to make altered states more accessible, especially creative states and meditative states. We see this same evolution also occurring in present scientific investigations into the electrophysiological profiles of meditation and advanced yogic states of consciousness (Van't Westeinde & Patel, 2022).

So, it should be no surprise to find clients integrating trauma and attempting to move to higher levels of integration beyond their basic clinical resolutions. The definition of clinical resolution is defined by academics but also by insurance companies, think DSM, as well as the individual clinician's training and experience. In reviewing the matter from a multidisciplinary perspective, we should expect trauma to be a fundamental and ubiquitous part of human experience and transformation, as well as expecting its initial resolution to propel individuals into higher states of integration and transformation, to learn from their experience. It is a part of individual growth and maturation.

Through NFB training the client will tacitly experience the power of the transformative process and the efficacy of self-regulation and self-management through neuromodulation. Consequently, clients will often be drawn to other new possible avenues of integration such as yoga, meditation, and prayer. They are likely at first to see these traditional methods as an economical means to further manage their residual grief and maintain their emotional balance. However, these practices also carry an invitation to a better way of experiencing life. Ultimately, this leads naturally to an existential shift that brings greater satisfaction and a more spiritual perspective. Resolving trauma challenges our existing beliefs and demonstrates there are alternative ways of being in the world and interpreting it. This often inspires further search with renewed interest in what is possible.

The five stages of grief have been outlined and well understood for at least a half century. They are key implicit components to the 12-step process. As previously discussed, the majority of those who drink heavily and develop alcoholism are self-medicating (Bradshaw, 1988). Often this is related to grief derived from trauma. This was well recognized long ago in the literature on alcoholism. Most of Bradshaw's work dealt with the underlying trauma associated with alcoholism. It is notable that a key component in 12-step is acceptance of a higher power. Anyone who studies or engages the method can tell you it is an absolutely critical component to the process. The serenity prayer, which is so central to 12-step, captures the essence of acceptance and the final stage of the five stages of grief.

In reviewing the literature on complex trauma, the ISTSS does not mention transpersonal solutions to trauma other than yoga and meditation and lists them as complementary, alternative, and integrative treatments. There is a paucity of research on the topic. This is ironic since it is so common for individuals to turn to spiritual and religious themes when dealing with disaster and when integrating PTSD as well as complex trauma. However, religion by definition cannot be considered an empirical solution in a scientific paradigm reviewing complex trauma. There is evidence for efficacy of yoga and meditation, but not mantra meditation (Forbes et al., 2020).

A key component of finding acceptance is finding meaning. In the subfield of sociology, known as dramaturgy, it is sometimes argued that human beings are meaning machines. Several prominent psychiatrists and neuroscientists have made similar arguments. In so many cases, and in my 20 years of clinical experience, I would say in most cases in my records, clients begin to identify transpersonal and spiritual themes early in their recovery and integration of past trauma. These themes are common and often appear to be an essential part of recovery. One of the first therapists to work with this perspective was the founder of Logotherapy, the renowned author, psychiatrist and clinician

Viktor Frankl (1962) which addresses man's search for meaning as a consequence of suffering. He proposed that man has a "will to meaning." Based on the fact that the "narrative" is a main vehicle for transmission and preservation of human tradition and related memories, this perspective carries significant weight. It agrees with many other authors and researchers from different disciplines as cited above.

So, there are many cross disciplinary threads with long traditions and sound arguments in this direction. The fundamental term WTF, is a basic expression of "why is this happening?" Its ubiquitous use in texting and over the internet is proof of its importance to human information processing. Most people dealing with trauma ask the question "why is this happening to me?" They look for a reason, an explanation, a meaning. To disregard this observation is like disregarding the existence of the taste of chocolate and the validity of qualia in the human experience. No good neuroscientist can ignore this fact.

The Counseling Community and Spirit

Religious and spiritual (RS) intervention has been found highly effective in over 32 studies and 16 of those found RS more effective than non-RS intervention (Corey, 2006). A more recent meta-analysis involving a combined N > 7,181 subjects found efficacy of RS resulted in statistically significant greater and spiritual improvement than non-RS treatments (Captari et al., 2018). Unfortunately, counselors typically interview clients about every aspect of their life but often do not ask about the role of spirituality or relation in their life (Post & Wade, 2009; Corey, 2006). Much of what follows is derived from Corey et al.'s findings. Corey argues,

"Spiritual and religious matters are therapeutically relevant, ethically appropriate, and potentially significant topics for the practice of counseling in secular settings. Counselors must be prepared to deal with their clients' issues of the human spirit."

Movement to address the sacred had significantly surfaced in the field of counseling by 2009. Religion and spirituality (RS) is an important source of meaning and identity for most people. It is a key reference point for framing the important events in people's lives. Major issues related to RS that can unknowingly confound clinical efforts include anger at God, existential doubts, or spiritual abuse. There has been growing empirical evidence that RS can contribute significantly to a client's health and psychological well-being. Contributing to these findings is research showing a positive relationship between religion and health. Most of the US population identifies as religious and spiritual. When and how to address these problems is the biggest question.

Religion is usually related to an institutional form of spirituality, which is a term about connectivity to the transcendent. In this metanalysis of the literature by Corey et al. involving empirical studies, both concepts of RS were used.

In 1985, 35% of psychologists based their life on religious beliefs compared to 72% of public. Yet 52% of psychologists indicated that spirituality was very important and 28% fairly important. In that period during the height of domination of the sciences by materialist reductionism in psychology, psychologists were five times more likely to deny belief in God. By 2007, however, as postmodernism settled into the mainstream, most psychologists were more likely to describe themselves as spiritual but not religious. They viewed client religiosity as positive and 82% believed it was beneficial to mental health.

Unfortunately, the farther from Christianity a client was, the more pathological their religiosity was perceived. Unfortunately, around that same time, only 13% of training directors and program leaders reported offering a specific course on religion/spirituality and psychology. This significantly undermined the impetus for clinicians to engage their clients having little information, tools, and resources.

In 2016 Pew Research reported that 89% of Americans believe in God or a Universal Spirit (Captari, et al., 2018). Eighty percent pray regularly and 50% belong to a house of worship. RS fosters increased social belonging, connection, and support. It aids through meaning-making, coping, and resilience. However, it can also be used as a defense against unresolved psychological conflicts.

"When people walk into the therapist's office," Pargament (2011), writes, "they do not leave their spirituality behind in the waiting room (p. 4). They bring their spiritual beliefs, practices, experiences, values, relationships, and struggles along with them."

These figures contrast with the fact only 35% of psychologists compared to 75% of public find their approach to life significantly influenced by RS. Some of these may be due to the fact that RS provides psychological support to the general public that counselors obtain through their own professional channels.

Those in the field who did engage in this aspect of counseling tended to be prudent enough to avoid imposing their own values on clients. These clinicians reported that RS issues tended to emerge gradually in the therapeutic relationship. This suggests that this is part of the natural sequence of the healing progression and agrees with the emerging model in AT training. It also reflects the clinical models emerging around complex trauma and the traditional AA model in that clients require basic work on uncovery, discovery, and integration first, before moving into more refined levels of integration. It is also in harmony with traditional yogic models of spiritual progression.

Clinicians in the past have tended to base their intervention on the RS background of their client and used meditation and scripture or prayer. The research suggests that clients were less comfortable with clinicians leading prayer instead of the clergy. They also found that gaining explicit client agreement to explore this area was crucial to engaging clients at this level of exchange and a critical component to the process. When a sample of religiously diverse clients were asked about willingness to discuss religious/spiritual (RS) issues, 55% wanted to do so and the other felt it was not relevant or, as mentioned, wanted to reserve that for clergy. In a survey of over 2000 clients from public and private universities across the US, 20% reported distress related to RS matters. This by itself should demonstrate the need for enhanced methods such as AT for addressing these issues.

Moderate to conservative Christians who were interviewed in various studies did not believe secular counselors would respect their beliefs, and this is an important impediment. Appropriate initiation utilizing respectful probes could overcome such obstacles. Most clients actually did expect prayer and scriptural references as most appropriate. They did tend to prefer Christian counselors, and in these instances, were more comfortable with, and expected in session, prayer. In these contexts, however, they did not like the idea of laying on of hands or encouragement to memorize scripture. Again, in some cases, there was discomfort with prayer together as they considered these ecclesiastical prerogatives and responsibilities. Latter Day Saints, however, were less likely to accept in-session prayer than other Christians.

Researchers tended to find being versed in some major tenants of a religion is helpful, and I found it very important to be at least basically familiar with major religious perspectives in dealing with my clients. However, even more important in dealing with these issues is openness. This was important for both affiliated and non-affiliated religious clients. Clients are most responsive to therapists who provided an environment of openness, acceptance, and safety, and especially respect. Consequently, most researchers reporting on RS found that clinicians need to thoroughly explore their own spiritual/religious experiences and biases, otherwise they are more likely to make poorer clinical judgements. Interventions where therapists taught spiritual concepts related to the client's situation were most effective. Meditation or religious imagery was less effective. Finally, everyone responds differently and uniquely to efforts to include RS in therapeutic settings.

Suffering and Meaning

Joseph Campbell proposed that one of the most fundamental explanations for the human experience was embedded in the Hero's Journey. By definition, the metaphor of the journey provides a meaning to what often otherwise is experienced as incomprehensible suffering. It argues that a search for an individual narrative is fundamental to individual identity and shared experience in social exchanges. It is the tender and currency of social exchange and shared experience that forms groups, families, tribes, and identity networks, even on the web.

Buddha's primary premise is that life is dukkha, or suffering, and his most famous contribution was a solution to suffering. It was an experiential solution and not a philosophical solution. One of the main aspects of this solution, but by no means the stand-alone solution, was meditation—now a popular therapy. Another key aspect, however, was the transactional aspect that occurred within the community of like-minded individuals, or Sangha, and the eight recommended ways to conduct yourself within that group as well as in the world in general. Although yoga had a similar aspect, it was far more philosophical and cerebral. Neurofeedback has great similarity to these meditational approaches in that they all generate fundamental changes in neural processing and dynamics, and even structure. Neurofeedback, however, is always in danger of being too self-reliant in terms of being the ultimate solution. This is why New Mind introduced the concept of functional neurofeedback to the field in general. This is done to include the other social and transactional components that are often a part of modern clinical psychology.

The work of Herbert Benson (2000) in identifying the common physiological features of prayer and meditation in his research at Harvard on stress found that the traditional methods of attaining important altered states and associated insights were key to helping people deal with stress, suffering, and trauma. Conflict and tragedy are a part of daily existence, as is the stress produced in securing the physiological, social, and psychological basics for life. The Buddhist psychology proposes a popular solution that has endured for thousands of years. Yet its core understanding is similar to Christianity according to one of the most famous contemporary Buddhist Zen masters (Hanh, 2007). This was also proposed by countless investigators and authors such as in the Perennial Philosophy (Huxley, 1946), philosophers like Alan Watts (1961), as well as scientists like Daniel Goleman (1988) and Richie Davidson (Davidson et al., 2003) who found significant psychological, physiological, and historical support as well. In fact, philosopher and author Alan Watts proposed before most in the mid-20th century in his book Psychotherapy East and West (1961) that both traditions were engaging in traditional psychotherapy. These names may now be unfamiliar, but they pioneered the convergence of eastern and western psychology and sensitized us to the validity of many traditional perspectives which had been impulsively cast aside without sufficient scientific grounds by the dominant materialist agenda of the 20th century.

Ultimately, individuals must reconcile themselves to what is occurring that is challenging their CNS on a regular basis. A key component of that must be a narrative because it is a key feature of one of the most present effective solutions to complex trauma, and that is CBT. At the same time, one of the primary investigators on anxiety, trauma, and the amygdala, Joseph LeDoux (2015) makes powerful laboratory research-based arguments that normalization of distressed network activity requires integration of implicit memory fragments into an explicit cortical narrative. We must also integrate with the group, identity group, or Sangha network externally as much as we must integrate with dissonant networks internally.

But as we have seen, cognition, emotion, and CNS activity are tightly interwoven, and in human beings, they are interdependent. The method of imaginal exposure in trauma recovery techniques that I have previously discussed confirms this fact. So, we can choose to recognize this component early in the treatment process or later, but sooner or later we must deal with it if integration is to be established at a high level. It is also critical for physical healing, and that insight is finally emerging in modern medicine as well.

Medical Practitioners and Spirit

Medical professionals, like psychological counseling clinicians, rarely take the spiritual dimension of healing into account (Lopez-Tarrida et al., 2021). The key reasons cited in the literature are similar to the counseling literature findings. The reasons cited were lack of training, lack of time, and fear about addressing this dimension of healing. This should not be surprising as it is not addressed in medical school, and not well addressed in counseling. Yet that may change in the future as research points out.

There is serious and progressive interest in the spiritual aspects of healing in physical medicine today. In July of 2022, as I write this book, a paper published In JAMA entitled "Spirituality in Serious Illness and Health" (Balboni et al., 2022) was hailed as one of the most important contributions to the literature on religion, spirituality, and health of the past 20 years. This was the result of a project at Harvard University supported by the John Templeton Foundation. They found strong evidence for the following:

1. Spirituality is important for most patients.
2. Spiritual needs are common.
3. Spiritual care is frequently desired by patients.
4. Spiritual needs are infrequently addressed in medical care.
5. Spirituality can play a role in medical decision-making.
6. Spiritual care is infrequent in medical care.

7. Unaddressed spiritual needs are associated with poorer patient QOL.

8. Provision of spiritual care is associated with better patient end-of-life (EOL) outcomes.

Based on these findings the top three implications emerged:

1. Routinely incorporate spiritual care into the medical care of patients with serious illness.

2. Include spiritual care education in the training of all members of the interdisciplinary medical team caring for the seriously ill.

3. Include specialty practitioners of spiritual care (e.g., chaplains) in the care of patients with serious illness.

The final consideration for counselors in all this from a professional perspective should be to consider the APA's position. The APA Ethics code (2017, Principle E) indicates that to ignore religion as an issue "may not only be unethical but lead to malpractice. It is an asset to the therapeutic process." It is important to understand what RS means to the clients and how it shapes their beliefs and behavior. Has it been a source of oppression or abuse? Do they have internal conflicts with it? How do they see it and use it as a source of strength?

The Client's Innate Inclination and Direction

As clients begin to heal from trauma, they frequently have spontaneous experiences that can only be articulated in psychology by applying the literature on research arising from transpersonal psychology and affiliated findings in neuroimaging and neurotheological theory. In addition, experts in consciousness research such as Christoff Koch (2019) and in the field of neurophysics such as Paul Nunez have also contributed significantly to a new understanding of the relationship between physics, consciousness, and human behavior and how they relate to human pain and suffering. These paradigms have begun gaining research momentum since the 1980s. They contain the vocabulary and the theory required to interpret and manage these categories of spontaneous experiences. To deny them with materialist reductionist arguments, that they should be rejected for consideration because they are not well supported in research, is anti-scientific in itself because there is considerable research and theory already in existence that supports their serious consideration (Taylor, 2018).

In reality, the reductionist agenda only supports a small, aggressive, and shrinking but very vocal segment of the scientific community. The materialistic and reductionist agenda they promote is derived from a misconstrued Newtonian perspective of energy and matter that does not withstand the litmus test of quantum mechanics. The majority of the founders of the quantum paradigm acknowledged that our popular materialist perspective was no longer valid, and the majority embraced a more spiritually based perspective of physics and reality (Selbie, 2021). From a psychological and sociological perspective, that should already make their motivations suspect and when the literature on the epistemology of science is considered their arguments fail to hold water (Soutar, 1996; Hoffman, 2019). Their message is clear and well documented, but to get immersed in the arguments in this area would at this point would only constitute a distraction not truly pertinent to this topic. Suffice it to say that the underlying spiritual component of human reality and experience is scientifically valid, documented, experimentally confirmed, and replicated. It has profound implications for human behavior and the sciences that claim to explain human behavior. Clinical psychology and counseling attempt to declare that their paradigms and methods are founded upon the behavioral sciences, and yet the behavioral sciences have failed to reconcile themselves with the quantum paradigm. This is a matter of human politics and not scientific validity.

The main reason for this discussion and review is to highlight the important and profound clinical implications for this oversight in clinical theory. The humanistic movement in psychology in the twentieth century, as represented by Maslow and others, attempted to compensate for this as it confronted the failure of scientific positivism in the form of behaviorism in psychology, and the agenda of Carnap in physics and the other so called "hard sciences" (Soutar, 1996). To disregard and minimize experiences that the client finds fundamental to their identity, efficacy, locus of control, as well as definitive of their relationships etc. would be an unsupportable strategy that contradicts the fundamental scientifically and clinically derived principles of effective therapy. Yet the fundamental spiritual dimensions of clients, of being human, are typically not accommodated but instead politely ignored.

It is a fact that, like consciousness, most people have transpersonal experiences, and they cannot be empirically quantified and measured in the same manner as physical phenomena. These experiences are considered by most people on the planet to be critical aspects of their lives. The evidence of brutal religious warfare going back thousands of years should be sufficiently significant to lend serious weight to this argument. Nowhere does it become more evident than in the transformations in personality and behavior that attend these experiences in the clinical setting because of doing trauma work. My clinical records and notes going back almost 30 years indicate that significantly increased spiritual and metaphysical concerns is a common outcome of AT training. It is also readily apparent that the level of transformation of personality as measured by both qEEG and psychometrics is highly correlated with the intensity of spiritual re-orientation and growth. That this is due to spontaneous and unprovoked psychophysiological responses to NFB requires a clinical response and protocol management strategy. Unfortunately, as I have noted and cited according to the research literature, the majority of therapists are neither trained nor prepared for this inevitability. In these situations, it is incumbent upon the therapist to refer to experts who are prepared for this situation.

Fortunately, there is presently a growing psychiatric and psychological technology emerging that specializes in this area involving psychedelics (Belouin & Henningfield, 2018). In addition, it is morally and ethically, if not practically, incumbent on those who engage in AT training to do their own training. BCIA acknowledges this even for the most basic certification in EEG biofeedback (BCIA Blueprint of Knowledge). For the majority of NFB practitioners, fulfillment of this requirement with respect to AT training directly confronts them with the spiritual dilemma, because they too have spontaneous experiences that must be processed, like it or not. In the long run denial is not an effective clinical option.

The Spiritual Component of AT

The transpersonal or spiritual component of the AT clinical experience proposes another level of integration beyond physical experience or social experience. Based on the New Mind group's data (over 500 clinics on average for over 10 years) as well as the peer reviewed literature, it appears that facilitation of the transpersonal experience provides important benefits that support critical positive changes in the client's personality and life experience moving forward. Accordingly, we have continued to develop and enhance this aspect of our AT training based upon our observations of this trend as well as ongoing feedback from the workshop attendees. We fully recognize that AT technology itself was derived from the metaphysical perspective of Elmer and Alice Green, and it is our intention to carry on that tradition. Elmer believed he was documenting and enhancing transpersonal phenomena through biofeedback measures (Green & Green, 1977). In addition, it should be noted that it is the 12-step component that was critical to the effectiveness of the AT methodology that both Peniston and eventually Scott successfully employed (Scott, 2017). A fundamental requirement of 12-step work is belief in a higher power and the research supporting the efficacy of 12-step, though endlessly controversial, I find abundant and persuasive. I have also found it a highly effective component in my 25 years of neurofeedback practice. As someone who began his research career in addictions and family systems, I have often observed its power to change people when nothing else could. In my graduate research I conducted 12-step study groups for several years, I can also attest to its efficacy.

Given the validity of our experience, arguments, and direct observations, how should we proceed to implement our understanding of these factors? In my book The Automatic Self, I provided a simple and basic framework for understanding these phenomena from a social psychological perspective. It was intentionally set forward in a simple format to provide access to my audience of interest, and not jaded academics and reductionist enthusiasts whose stalwart bias would automatically exclude them. The inspiration for this perspective was derived from Buddhist psychology as much as from western social psychology, especially the sub-disciplines of symbolic interactionism and dramaturgy. The legitimacy of the science behind the perspective was clear to me based on insights gained doing my Ph.D. dissertation (Soutar, 1996) on the implications of quantum mechanics for methods in the behavioral sciences. Those interested in the academic details are referred to my 8 hours of lecture in the videos posted on the internet as well as my dissertation (Soutar, 1996), apparently available on the internet.

My fundamental proposal is that like everything else in biology, we are an aggregate of routines which we forge over the successive stages of our life experiences. The manifestation of those routines arises from patterns of social interaction in the course of our biological and social development. Those routines inform and shape our physical bodies including our brain and central nervous system. The growth and development of the human brain profoundly impacts its structure, network development, and efficiency. Structural discontinuities are highly correlated with mental disorders and behavioral problems (Menon, 2011). They are key components of network processing at all levels including the small world structure of the human connectome. That consequential shaping of our fundamental neuronal infrastructure profoundly influences our psychophysiological patterns and feeds back into our ongoing experience of social and physical reality. The ACE study proves that this process is so profound that it defines and predicts our mental and physical health over the course of our lifetime. As symbolic interactionism points out, life is problematic, and we are in a constant state of negotiation (Blumer, 1969). For many of us, those negotiations do not turn out well, not only because of poorly functioning neurology but also because we have been poorly trained for negotiation throughout our developmental phases. Our social accuracy (Fiske & Taylor, 1991) is poor and we are not able to access the social resources we need to thrive (Soutar, 2008). Those interested in the details of this perspective can find it and related statistical analysis in my paper on the construction of the Interactive Self Inventory (Soutar, 2008) which was designed to measure social accuracy.

Those with poor social accuracy have received an inheritance of genetics that bias them in this direction. The now well-known and replicated concordance studies done in the 20th century demonstrated that this inheritance can be significantly transcended through appropriate socialization. Those with a poor introduction to social reality and the burden of transgenerational behavior patterns in conjunction with genetic polymorphisms do not fare well as they become physically ill and earn the label of mentally disordered. Their experience is deeply burdensome and profoundly overwhelming. Their suffering is ongoing and intense and mostly invisible to those who have not shared in that experience. Clinicians who have not come from that background of suffering and have not significantly confronted it are, in a sense, often doomed to mediocrity because they will never fully comprehend their clients or the process of recovery from their dark socialization journey. They will not be able to walk the walk and talk the talk and their clients will know it at a fundamental level. Authenticity in this realm, unlike the realm of human politics, cannot be faked.

In this socialization into profound suffering, the human CNS becomes fractured, and the major networks of the brain become disorganized. We have previously reviewed the powerful destructive impact of excitotoxicity, due to emotional trauma, and sleep deprivation alone on the brain. This disorganization leads to a profound loss of ATP efficiency as well as degradation of neuronal tissue due to the associated chronic inflammatory process. Those suffering become captured by their neuronally isolating self-referential default mode network and imprisoned by a hyperactive amygdala and affiliated subcortical networks. Their hemispheric symmetry shifts as they progressively move into a withdrawal pattern of behavior. Anger and fear dominate their fundamental emotional landscape as they cannot find anyone to trust, and their locus of control moves outward and they experience themselves as victims of those more powerful. The pathways to adapt to this fatalistic and grim scenario are as infinite as humans are complex, and this is reflected in their behavior, their physiology, and their brainwave patterns. We can, however, categorize aspects of the process in a limited fashion if we don't succumb to simplistic symptomatic lists such as found in the DSM.

I have outlined the psychological, sociological, physiological, and neurological patterns of this socialization into living purgatory and hell as my clients have described it in the previous chapters, as well as explicated how to weave the golden thread of NFB training to guide them out of the labyrinth of this Minotaur's dark and painful realm. The experience of the hopelessness of profound depression or the terror of derealization and depersonalization associated with severe panic attacks is impossible to even begin to communicate with others, much less the horror of dark hallucinations. Those clinicians who have experienced the effects of psychotomimetic compounds and worked on their own recovery can begin to appreciate the extremes of this dark side of the human psyche. Instead, we speak of these experiences of our clients in antiseptic terms to protect ourselves and maintain control but distance us from insight and solutions.

As clients engage in AT work, they are operantly conditioned to move into a liminal state that bifurcates their fully alert conscious experience. The experiencer and the experienced begin to separate in the crossover state. There is that which engages in the activities forged around suffering, and that which observes and watches without judgement or comment. Clients begin to become aware of another aspect of themselves that is not apparently biographical. In his book The Feeling of What Happens, Antonio Damasio (1999) describes this fundamental non-biographical consciousness as arising from the trigeminal plane of the brainstem, and filtering up though the limbic and self-referential systems of the cortex like light through a colorful pane glass window of personal history. At the same time, the research of Joseph LeDoux (2015) explains to us how implicit fragments of trauma memory must aggregate and become integrated into a cortical narrative in order for the biological drive to neurologically integrate for optimal functioning diminishes. All of this points to the evidence that we as individuals are an ongoing narrative always under construction and typically overidentified and lost in that narrative. Oddly, this jibes with ancient Mundaka Upanishadic commentary that there are two birds on the tree of life, and one quietly watches as the other eats of its fruit. One is in the world, and one is in the world but not of it.

You may have noticed that we have discussed this concept previously from another angle in considering methods of integration of trauma and exposure therapy. The ironic outcome of effective trauma therapy is that it amplifies this theater of the viewer and the viewed, and the canvas of conscious experience becomes somewhat separate and amplified so that the experience and the experiencer become more salient. It is this movement into the observer that the quiet place of integration noted by van der Kolk and everyone else seriously working with trauma that allows integration of trauma. But what is not highly noted is the powerful insights that follow for the individual regarding their identity and relationship to life itself. We are now in the realm of Rollo Therapy of Rollo May, the work of Victor Fankl and the human drive for meaning. Clients begin to realize that they are more than their thoughts and their experiences. They can't help but begin to often notice an intense loving presence in those healing moments and that this presence is the catalyst for their integration. Jung would argue that this presence is their Core Self, the Hindus would argue it is the Atman and essentially

Brahma himself while the Buddhists would say it is Buddha, the Christians Christ, and the Muslims simply Allah. This type of experience is especially powerfully salient in psychedelic therapy utilizing serotonergic psychedelics (Thal et al., 2021). The unique EEG signatures of these states are well documented at this point in time (Tagliazucchi, 2020) as are the signatures of meditative states. Those who I have done AT training with, who have experienced contemporaneous psychiatric ketamine work, commonly have this type of experience but do not always understand the implications of their experience until they have someone present who has been through a similar experience and beyond and can guide them to the appropriate insights. It is critical to have direct tacit experience to guide others.

The outcome of many integrative experiences is a growing interest in transpersonal and spiritual experience. The majority of my clients who achieve significant integration begin to seek out sources that can satisfy their thirst for explanations and guidance in these areas. For many, a return to their original religious background can suffice at first, but soon they may find the intensity of their personal integration experience has moved them beyond the basic religious experience of their upbringing. They may find that others who have not worked through their own traumas lack a shared depth of experience that generates an authenticity that the client craves. Some clients grasp this intuitively and immediately and seek alternative routes. This can be frustrating, as many supposed teachers and groups thrive off of co-dependent followers and seekers, as well as those with partially resolved traumas searching for an alternative route of integration. It is easy to get lost in the spiritual materialism (Trungpa, 2002) of the spiritual marketplace especially with the internet and its Babbeling (spelling intentional) cacophony.

Having a clinician who has earned their trust and, in many cases, unavoidably introduced them to an alternative construction of their reality that they are unsure of how to manage could be confusing and frustrating. I have found that there is often a great sense of relief and continuity for most clients in moving forward to explore this side of their potential. It is my clinical experience that this next step is crucial to begin resolving the much deeper wounds of trauma that continue to endure despite every therapeutic effort at full resolution. As we have noted in the research sections of the book on complex trauma, there is presently no therapeutic technique that can promise complete resolution of complex trauma. It can be well argued that it is only by engaging the transpersonal realm of experience on a regular basis that higher levels of integration can take place. Ironically, those who are significantly traumatized have the greatest opportunity to discover and explore this aspect of human experience. In a sense their pain is a gift and an invitation to find more wisdom and integration and fulfillment than perhaps another individual who is less challenged in their socialization and developmental process. In this case the first will be the last. As Judy Lubar used to say, "You've made a down payment on wisdom, now it's time to collect."

The Process

This is where Adam Crane and I came together in our long-term vision for NFB. It was clear to us from our own experience that AT can easily lead into profound altered states and transpersonal experiences, and that those too could be amplified and refined through the NFB process and so Adam referred to it as The Process. He came up with the term profound attention for the experience of accessing what Daniel Goleman (1988) called the access state. In Yogic tradition this is called Samadhi. This is the state where observer and observed have first merged and then separated in a novel fashion at a higher level of conscious processing. The shift is from the bird eating the fruit to the bird quietly watching in the Upanishadic metaphor previously mentioned. The witness state, in this approach, is enhanced through NFB by training alpha synchrony (and not the intermediary gamma frequency). Practicing these techniques typically leads to an experience of profound attention if the client is properly prepared. Les Femi (1978), who was running experiments at Princeton, had done experiments demonstrating this to be the case, and TM University had done research to show the same thing (Banquet, 1973).

In true meditational states, we are seeking neural quiescence so that the fundamental consciousness can shine through without the stained-glass colors of personal narrative. But this is not a gimmick or quick trick. It takes self-discipline and practice. NFB, however, helps clients identify when they are accurately on track to achieving the access state through the right effort of profound attention. Their journey will be landmarked by growing and more frequent spontaneous experiences of joy and bliss that amplify their association with their Core Self and the transcendent aspect of their daily experience. They move powerfully into an experience of the merging of their core being and consciousness into the profound universal present through the exercise of profound attention. The precision of brainwave feedback can significantly accelerate the learning process associated with this undertaking. It provides guidance and support and the kind that 21st century technologically socialized individuals require. Without it, most individuals in our culture will fail at meditation and its associated states. Our way is the way of technology—just look around.

The Method

The goal in The Process, once the client has been through fast wave training for resilience, and AT training for integration, is to engage in a specific type of alpha synchrony training. Again, this technique is useless to engage in until the other steps are taken, and can otherwise only end in either failure or aggravated mental disorder. This is not speculation or from our data only. This is consistent with a more ancient perspective, think Yogic Tapas, confirmed by contemporary research. In addition, this training will not work particularly well until the brainwave distribution has been significantly normalized. The reason for this would only be clear to those who have extensive experience in the field. The brain reorganizes in stages or iterations that often mimic evolving heteroclinic attractor patterns (Miller, 2016). This is predictive of the patterns we observe in New Mind pre-post mapping sequences, wherein map patterns over time appear to repeat in a progressive reorganizational pattern, while at the same time displaying a progressively normalizing pattern with respect to the neurometric norms of the database system.

Observing only one neurometric dimension blinds clinicians to the more subtle patterns across neurometric dimensions that reflect an interdependent system. A unidirectional change in alpha amplitude may reflect a series of underlying changes occurring in other dimensions that are overlooked. For instance, the brain must change phase and coherence patterns in specific sequences to support the changes in magnitude. The changes go hand in hand, system wide, at all levels of accommodation. The same requirements apply to dominant frequency and asymmetry. Again, this is a consequence of the systemic nature of brain function.

The high amplitude alpha of PTSD is not the same as the high alpha amplitude of meditation. They have profoundly different underlying patterns that are radically different. Effective meditation must be arrived at through a preamble of integrative iterations. The brain cannot fully integrate with the rest of the body until trauma is significantly integrated and the brain itself cannot support the movement into these altered states without a higher level of neuronal efficiency. The clinical preamble to this is the modern version of the ancient Tapas required to prepare for years of meditational practice.

There are many people who have come to my office who have been meditating for years but have been somewhat concerned about their progress. They are usually looking for a brainmap to confirm their progress. Unfortunately, in the majority of the cases, I have found that most do not have the signature of an advanced meditator when I am mapping them during their meditation process. I expect that this is because they have not been properly taught or are do it yourselfers (DIY), and indeed many admit to that category. As simple as meditation is in many respects, it is very difficult for westerners to do. It is deceptively simple, as we have said, because it requires a highly disciplined neurological preamble of integration. We bring too much complexity and impatience to the process of meditation. There are a lot of stereotypes and myths surrounding the practice that led to misconceptions regarding its goals and purpose.

The primary purpose of meditation is not to arrive at an endless ecstatic state of omniscience and omnipotence. It would be impossible to even live long in such a state. That is not to say that ecstatic states and bliss are not at times artifacts of the process, because they are indeed, but they are not the point. Neither is the purpose of meditation to attain supernatural powers, as Patanjali points out in his 2500-year-old treatise on the topic (Isherwood & Prabhavananda, 1969). The point of meditation is rather to obtain a fundamental shift in perspective based on insight that is so profound that it leads to an experience of completeness, connectivity, and fulfillment that is ongoing. This experience is punctuated by periodic experiences of intense ecstasy, but these are not the norm. They are merely moments of peak CNS resonance and integration, and act as resets in an environment which is often aversive and stressful to the organism as a whole. The fundamental insights shift the mind and the organism into a new level of functioning.

Of further importance is the fact that, technically, integration never ends because of this dialectic between the observer and the observed and between the organism and its social and ecological environment. Learning, innovation, and adaptation are necessary ongoing functions of life on this planet and in this dimension. A key aspect of REM sleep is to integrate daily learning into the brain's functional and behavioral repertoire. There is no end point to reach other than dissolution of the experiential matrix of this aspect of the time-space dimension. Continuation in other dimensions can have little meaningful translation to this dimension. The language and data processing would be unique to that environment just as it is here in this dimension.

The Process leads to an understanding through direct experience of the aspect of reunification of the self and the whole beyond the verbal realm of conception, which is still a social artifact. Like the physicist trying to understand through physical experiment the fundamental nature of gravity, the spiritualist seeks to understand the fundamental nature of unity. As the renowned physicist David Bohm explains it, everything is connected to everything else, and the universe is indeed holographic in nature. Indeed, the unified field theory and the completion of quantum mechanics seeks a mathematical experience that parallels the spiritual journey. One is the fulfillment of a quantitative agenda and the other is the fulfillment of a qualitative agenda. The understanding of the physicist relies upon the language of mathematics. The understanding of the spiritualist relies upon the more empirical and tacit language of direct perception through the experience of qualia. In fact, the spiritual student expects to go beyond even the percept until the nature of pure consciousness is experienced as the source of unity itself.

There is in fact a scientific basis for this perspective of consciousness as the fundamental unifying factor of the universe, and this is found in quantum mechanics (Koch, 2019). Monastic idealism emerging from the framework of quantum mechanics provides a basic theoretical perspective and mathematics for explaining both consciousness and matter (Goswami, 1993). The theories of quantum entanglement explain how consciousness itself divides into a self-reflective process that elicits the material perceptual experience from the implicate order. As the Buddhist doctrine of dependent origination predicts, subject and object co-arise. This is further articulated in the concept and experience of interbeing (Hanh, 2007). The wave form collapses whenever and wherever consciousness appears in the time-space continuum. This is no longer just speculation. The math has been done. Experiments have been done. The replications of Aspect's experiments, as reported by Goswami (1993) demonstrating the non-locality of consciousness have been replicated many times. It is interesting to watch many scientists try to deny the proofs of their own methodology when they don't like the conclusions (Bohm, 1980).

There are further evolving mathematical proofs regarding the nature of reality which indicate that it is iconic in nature (Hoffman, 2019). What is meant by that conceptualization is that consciousness accesses reality through a collapse of the wave function by focusing on perceptual configurations derived from the decomposition of the field of experience in the time-space matrix. This is analogous to clicking on icons on a desktop. The icons are not the thing in itself, but rather a means of accessing functions that explicate the thing-experience. Consciousness can then navigate the time-space matrix through this method using mind, which extends beyond the local time-space matrix into future and past. Mind transcends the flow of time-space experience but only up to a point where it progressively dissolves back into consciousness itself. Mind then is consciousness iterating itself through infinite feedback. If this is vague and overwhelming, then I refer you to Hoffman's work as well as Joseph Selbie (2021) who is more basic in his explanation.

The fundamental nature of our consciousness itself, from this perspective, is to iterate itself back into unity which almost becomes its definition. For human beings, this becomes a fundamental drive to process and integrate and to become highly efficient at that function in this time-space dimension. We are compelled to experience, learn, and integrate at the most basic level. Creativity is at our core, and it is fundamental to all life. Significant trauma disrupts and blocks that process, whether it is physical, neurological, or mental. In a sense, trauma is input that exceeds our ability to process output. A kink in the system if you will. In all systems this leads to entropy, which in our dimension ultimately leads to black holes or death.

Most clinicians will likely feel it is outside their scope of practice to venture into assisting their clients with their efforts to focus on their spiritual grow and recognize the limitations of making recommendations out of their own religious bias and affiliations. Yet many of those clients emerging are susceptible to cults because of their trauma background, or may even be survivor of cults, or worse, abuse at the hand of a major religious organization such as the Catholic Church. This calls on therapists to at least educate their clients on the pitfalls of seeking a group of likeminded individuals engaged in a spiritual journey, and help them establish guidelines in their search for spiritual affiliation. This dilemma has been recognized since the 1990s in publications such as Psychology Today as well as professional journals going back 20 years and we will review this shortly.

There are some basic generic educational steps therapists can take to help guide clients. Therapists can point out the client's particular vulnerabilities based on their specific history of abuse. They can direct clients to videos and publications that can assist them in recognizing bona fide groups and organizations and how to identify them. Clinicians can also educate clients regarding red flags they may encounter in their journey and how to extract themselves from compromising situations.

In many cases clients begin to identify spiritual themes early in their recovery and integration of past trauma. These themes are common and often appear to be an essential part of recovery. One of the first therapists to work with this perspective was the founder of Logotherapy, Viktor Frankl (1962) which addresses man's search for meaning as a consequence of suffering. He proposed that man has a "will to meaning." Based on the fact that the "narrative" is a main vehicle for transmission and preservation of human tradition and related memories, this perspective carries significant weight.

Modalities of Clinical Spiritual Processing

Although meditation was originally considered primarily an eastern religious tradition, mindfulness has come to be used at most of the major hospitals in the US. It is recommended for a large variety of physical health issues and for healing in general. It is also recommended by ISTSS as an important component of CPTSD recovery and healing. Much of this has to do with the research, efforts, programs and publications of Herbert Benson, Jon Kabat-Zinn and Richard Davidson. These features of meditation provide a legitimate opportunity for therapists to make recommendations regarding meditation practice as a starting point for clients. It should be noted, however, that meditation should not be engaged until a client has had significant neurofeedback training and trauma integration, as research in the past has shown it can precipitate traumatic feelings and memories that may overwhelm clients if they have not done the correct preparation for it.

Although prayer tends to be characterized by recommended methods and content depending on the religious context it is employed in, it does have some fundamental common features across religions, and evolved in the AA movement into the concept of the "higher power" and the Serenity Prayer. Within the AA community there was historically an intense debate of whether there should be any theistic references or resources employed. Eventually the concept of a "Higher Power' was determined to be crucial to the process of recovery. To date, this is still one of the most effective approaches to recovery from addiction and is well known and understood in the counseling literature and the counseling community. This in itself is a powerful testimony to the importance of the spiritual component as a major driving force in the alcoholic experience is self-medication to cope with the emotional and physical developmental trauma almost universally reported by recovering alcoholics. The Serenity Prayer is used by AA members from many different religious backgrounds to aid in the process of recovery and is considered to have very important and significant therapeutic value in many clinical settings.

In addition to the foregoing, there are also the extensive clinical theories and methods utilized by Carl Jung that are also employed in psychodynamic methods of clinical psychology. This includes the recognition of a core higher self and personal self that are found in many eastern psychological traditions. Jung actively recognized dream interpretation as well as lucid dreaming as effective clinical tools.

Lucid dreaming has become more recognized in the therapeutic setting as a valid method of integration since it has been validated in the laboratory as a very real and not uncommon aspect of human experience by Steven LaBerge (LaBerge & Rheingold, 1990). It is also a common component of traditional spiritual training with strong roots in Greek and Egyptian healing traditions. It is also a component of Tibetan Yogic tradition as well.

Visualization has also become another common method used in the modern clinical setting to assist clients in psychological healing. Visualization however, as previously mentioned, is a traditional tool in primitive and shamanistic religions and a key component of these spiritual approaches (Knoll, 1985).

In view of these clinical developments over the last century, it should be self-evident that psychology has progressively harnessed a significant portion of what was once considered the domain of religious practices to enhance clinical outcomes. There are developed research literature traditions for each one of these modalities with a variety of recommended methods of implementation. This provides the clinician with a broad array of tools to assist clients in integration and at the same time provide a seamless transition into more spiritually oriented practices that they can continue to develop after maturating beyond the clinical setting.

We have incorporated all of these methods in the New Mind AT methodology, as you may have noticed after reading the chapter on AT methods, and we present them in more detail in our AT workshops. Each of these techniques or methods provides effective changes and enduring insights for clients that aid significantly in their healing and further personal development. It provides them with authentic experiences in each methodological domain that affords them a valuable reference for assessing their encounters with these practices in other settings designated as primarily spiritual or religious in nature. It also assists them in experiencing the intrinsic value of these practices and methods based on their tacit experience rather than just hearsay or recommendations from friends and family. The transformative experiences they encountered extended legitimacy where there was likely doubt and reservations, which is why AA was inclined to use terms like higher power.

With these tools AT training has become one of the most powerful clinical tools we have developed. It has profoundly changed the lives of thousands upon thousands and provided them with a degree of healing they could not find elsewhere. It is the technological path to wisdom and insight. Whether it continues to do so depends on the continued evolution of the technology by the thousands of clinicians worldwide who have learned to use it and presently employ it.

References

Ainsworth, S., & Bowlby, J. (1991). An ethological approach to personality development. American Psychologist, 46, 331-341. http://dx.doi.org/10.1037/0003-066X.46.4.333

Alkadhi, K. (2013). Brain physiology and pathophysiology in mental stress. International Scholarly Research Notices Physiology, 2, 1-23. http://dx.doi.org/10.1155/2013/806104

Alves, P. N., Foulon, C., Karolis, V., Bzdok, D., Margulies, D. S., Volle, E., & Thiebaut de Schotten, M. (2019). An improved neuroanatomical model of the default-mode network reconciles previous neuroimaging and neuropathological findings. Communications Biology, 2, 370. https://doi.org/10.1038/s42003-019-0611-3

An, L., Wirth, U., Koch, D., Schirren, M., Drefs, M., Koliogiannis, D., Nieß, H., Andrassy, J., Guba, M., Bazhin, A. V., Werner, J., & Kühn, F. (2022). The role of gut-derived lipopolysaccharides and the intestinal barrier in fatty liver diseases. Journal of Gastrointestinal Surgery, 26(3), 671-683. https://doi.org/10.1007/s11605-021-05188-7

Anand, B. K., Chhina, G. S., & Singh, B. (1961). Some aspects of electroencephalographic studies in Yogis. Electroencephalography and Clinical Neurophysiology, 13, 452-456. https://doi.org/10.1016/0013-4694(61)90015-3

Andrillon, T., Poulsen, A. T., Hansen, L. K., Léger, D., & Kouider, S. (2016). Neural markers of responsiveness to the environment in human sleep. The Journal of Neuroscience, 36(24), 6583-6596. https://doi.org/10.1523/JNEUROSCI.0902-16.2016

Arns, M., Conners, C. K., & Kraemer, H. C. (2013). A decade of EEG theta/beta ratio research in ADHD: A meta-analysis. Journal of Attention Disorders, 17(5), 374-383. https://doi.org/10.1177/1087054712460087

Austin, J. H. (1998). Zen and the brain: Toward an understanding of meditation and consciousness. Cambridge, MA: The MIT Press.

Avram, J., Balteş, F. R., Miclea, M., & Miu, A. C. (2010). Frontal EEG activation asymmetry reflects cognitive biases in anxiety: Evidence from an emotional face Stroop task. Applied Psychophysiology and

Biofeedback, 35(4), 285-292. https://doi.org/10.1007/s10484-010-9138-6

Balboni, T. A., VanderWeele, T. J., Doan-Soares, S. D., Long, K. N. G., Ferrell, B. R., Fitchett, G., Koenig, H. G., Bain, P. A., Puchalski, C., Steinhauser, K. E., Sulmasy, D. P., & Koh, H. K. (2022). Spirituality in serious illness and health. JAMA, 328(2), 184–197. https://doi.org/10.1001/jama.2022.11086

Bandura, A. (1997). Self-efficacy: The exercise of control. W. H. Freeman/ Times Books/ Henry Holt & Co.

Banquet, J. P. (1973). Spectral analysis of the EEG in meditation. Electroencephalography and Clinical Neurophysiology, 35(2), 143-151. https://doi.org/10.1016/0013-4694(73)90170-3

Barnett, A. J., Man, V., & McAndrews, M. P. (2019). Parcellation of the hippocampus using resting functional connectivity in temporal lobe epilepsy. Frontiers in Neurology, 10, 920. https://doi.org/10.3389/fneur.2019.00920

Barry, J. (2018). The great influenza: The story of the deadliest pandemic in history. New York: Penguin Books.

Beck, J. S. (1995). Cognitive therapy: Basic and beyond. New York: The Guilford Press.

Belouin, S. J., & Henningfield, J. E. (2018). Psychedelics: Where we are now, why we got here, what we must do. Neuropharmacology, 142, 7-19. https://doi.org/10.1016/j.neuropharm.2018.02.018

Belujon, P., & Grace, A. A. (2017). Dopamine system dysregulation in major depressive disorders. The International Journal of Neuropsychopharmacology, 20(12), 1036-1046. https://doi.org/10.1093/ijnp/pyx056

Benson, H. (2000). The relaxation response. New York: Avon Books.

Birbaumer, N., Ruiz, S., & Sitaram, R. (2013). Learned regulation of brain metabolism. Trends in Cognitive Sciences, 17(6), 295-302. https://doi.org/10.1016/j.tics.2013.04.009

Blumer, H. (1969). Symbolic interactionism: Perspective and method.

Englewood Cliffs, NJ: Prentice-Hall, Inc.

Bohm, D. (1995). Wholeness and the implicate order. New York: Routledge.

Bolton, T. A. W., Wotruba, D., Buechler, R., Theodoridou, A., Michels, L., Kollias, S., Rössler, W., Heekeren, K., & Van De Ville, D. (2020). Triple network model dynamically revisited: Lower salience network state switching in pre-psychosis. Frontiers in Physiology, 11, 66. https://doi.org/10.3389/fphys.2020.00066

Bowlby, J. (1982). Attachment and loss (2nd ed.). New York: Basic Books.

Bradshaw, J. (1988). Bradshaw on the family: A revolutionary way of self discovery. Deerfield Beach, FL: Health Communications.

Brady, K. T., Killeen, T. K., Brewerton, T., & Lucerini, S. (2000). Comorbidity of psychiatric disorders and posttraumatic stress disorder. The Journal of Clinical Psychiatry, 61(Suppl 7), 22-32.

Broughton, R., & Hasan, J. (1995). Quantitative topographic electroencephalographic mapping during drowsiness and sleep onset. Journal of Clinical Neurophysiology, 12(4), 372-386.

Buckner, R. L., Andrews-Hanna, J. R., & Schacter, D. L. (2008). The brain's default network: Anatomy, function, and relevance to disease. Annals of the New York Academy of Sciences, 1124, 1-38. https://doi.org/10.1196/annals.1440.011

Budzynski, T. H. (1999). From EEG To neurofeedback. In James R. Evans & Andrew Abarbanel (Eds.), Quantitative EEG and neurofeedback (pp. 66-76). New York: Academic Press.

Buzsaki, G. (2006). Rhythms of the brain. New York: Oxford University Press.

Bzdok, D., Langner, R., Schilbach, L., Engemann, D. A., Laird, A. R., Fox, P. T., & Eickhoff, S. B. (2013). Segregation of the human medial prefrontal cortex in social cognition. Frontiers in Human Neuroscience, 7, 232. https://doi.org/10.3389/fnhum.2013.00232

Cade, M., & Coxhead, N. (1987). The awakened mind. Dorset, England: Element Books.

Callaghan, B. L., Fields, A., Gee, D. G., Gabard-Durnam, L., Caldera,

C., Humphreys, K. L., Goff, B., Flannery, J., Telzer, E. H., Shapiro, M., & Tottenham, N. (2020). Mind and gut: Associations between mood and gastrointestinal distress in children exposed to adversity. Development and Psychopathology, 32(1), 309-328. https://doi.org/10.1017/S0954579419000087

Callaway, T. G., & Bodenhamer-Davis, E. (2008). Long-term follow-up of a clinical replication of the Peniston protocol for chemical dependency. Journal of Neurotherapy, 12(4), 243-259. https://doi.org/10.1080/10874200802502060.

Campbell, J. (2008). The hero with a thousand faces (3rd edition). Novato, California: New World Library.

Captari, L. E., Hook, J. N., Hoyt, W., Davis, D. E., McElroy-Heltzel, S. E., & Worthington, E. L., Jr. (2018). Integrating clients' religion and spirituality within psychotherapy: A comprehensive meta-analysis. Journal of Clinical Psychology, 74(11), 1938-1951. https://doi.org/10.1002/jclp.22681

Chiovato, L., Magri, F., & Carlé, A. (2019). Hypothyroidism in context: Where we've been and where we're going. Advances in Therapy, 36(Suppl 2), 47-58. https://doi.org/10.1007/s12325-019-01080-8

Choi, S. W., Chi, S. E., Chung, S. Y., Kim, J. W., Ahn, C. Y., & Kim, H. T. (2011). Is alpha wave neurofeedback effective with randomized clinical trials in depression? A pilot study. Neuropsychobiology, 63(1), 43-51. https://doi.org/10.1159/000322290

Chu, B., Marwaha, K., Sanvictores, T., & Ayers, D. (2022). Physiology, stress reaction. In StatPearls. StatPearls Publishing.

Corey, G. (2006). Integrating spirituality in counseling practice. Vistas Online 2006. American Counseling Association. https://www.counseling.org/knowledge-center/vistas/by-year2/vistas-2006/docs/default-source/vistas/integrating-spirituality-in-counseling-practice#

Correia, A. S., & Vale, N. (2022). Tryptophan metabolism in depression: A narrative review with a focus on serotonin and kynurenine pathways. International Journal of Molecular Sciences, 23(15), 8493.

https://doi.org/10.3390/ijms23158493

Cowdin, N., Kobayashi, I., & Mellman, T. A. (2014). Theta frequency activity during rapid eye movement (REM) sleep is greater in people with resilience versus PTSD. Experimental Brain Research, 232(5), 1479-1485. https://doi.org/10.1007/s00221-014-3857-5

Cowen, P. J., & Browning, M. (2015). What has serotonin to do with depression?. World Psychiatry (WPA), 14(2), 158-160. https://doi.org/10.1002/wps.20229

Craig, A. D. (2005). Forebrain emotional asymmetry: A neuroanatomical basis?. Trends in Cognitive Sciences, 9(12), 566-571. https://doi.org/10.1016/j.tics.2005.10.005

Crane, A., & Soutar, R. (2000). Mindfitness training: Neurofeedback and the process. New York, NY: iUniverse.

Crocq, M. A., & Crocq, L. (2000). From shell shock and war neurosis to posttraumatic stress disorder: A history of psychotraumatology. Dialogues in Clinical Neuroscience, 2(1), 47-55. https://doi.org/10.31887/DCNS.2000.2.1/macrocq

Crosson, B. (1992). Subcortical functions in language and memory. NY: The Guilford Press.

Damasio, A. (1994). Descartes' error: Emotion, reason, and the human brain. New York, NY: Avon Books.

Damasio, A. (1999). The feeling of what happens: Body and emotion in the making of consciousness. New York: Harcourt Brace & Co.

Davidson, R. J. (2000). Affective style, psychopathology, and resilience: Brain mechanisms and plasticity. The American Psychologist, 55(11), 1196-1214. https://doi.org/10.1037//0003-066x.55.11.1196

Davidson, R. J., Kabat-Zinn, J., Schumacher, J., Rosenkranz, M., Muller, D., Santorelli, S. F., Urbanowski, F., Harrington, A., Bonus, K., & Sheridan, J. F. (2003). Alterations in brain and immune function produced by mindfulness meditation. Psychosomatic Medicine, 65(4), 564-570. https://doi.org/10.1097/01.psy.0000077505.67574.e3

Dehghani-Arani, F., Rostami, R., & Nadali, H. (2013). Neurofeedback training for opiate addiction: Improvement of mental health and craving. Applied Psychophysiology and Biofeedback, 38(2), 133-141. https://doi.org/10.1007/s10484-013-9218-5

Del Colle, A., Israelyan, N., & Gross Margolis, K. (2020). Novel aspects of enteric serotonergic signaling in health and brain-gut disease. American Journal of Physiology. Gastrointestinal and Liver Physiology, 318(1), G130-G143. https://doi.org/10.1152/ajpgi.00173.2019

Dietzel, I. D., Mohanasundaram, S., Niederkinkhaus, V., Hoffmann, G., Meyer, J. W., Reiners, C., Blasl, C., & Bohr, K. (2012). Thyroid hormone effects on sensory perception, mental speed, neuronal excitability and ion channel regulation. In N. K. Agrawal (Ed.), Thyroid hormone (pp. 85-122). IntechOpen. https://doi.org/10.5772/48310

Dunn, B. R., Hartigan, J. A., & Mikulas, W. L. (1999). Concentration and mindfulness meditations: Unique forms of consciousness?. Applied Psychophysiology and Biofeedback, 24(3), 147-165. https://doi.org/10.1023/a:1023498629385

Egner, T., & Gruzelier, J. H. (2003). Ecological validity of neurofeedback: Modulation of slow wave EEG enhances musical performance. Neuroreport, 14(9), 1221-1224. https://doi.org/10.1097/01.wnr.0000081875.45938.d1

Egner, T., Strawson, E., & Gruzelier, J. H. (2002). EEG signature and phenomenology of alpha/theta neurofeedback training versus mock feedback. Applied Psychophysiology and Biofeedback, 27(4), 261-270. https://doi.org/10.1023/a:1021063416558

Engels, A. S., Heller, W., Mohanty, A., Herrington, J. D., Banich, M. T., Webb, A. G., & Miller, G. A. (2007). Specificity of regional brain activity in anxiety types during emotion processing. Psychophysiology, 44(3), 352-363. https://doi.org/10.1111/j.1469-8986.2007.00518.x

Erlacher, D., Schädlich, M., Stumbrys, T., & Schredl, M. (2014). Time for actions in lucid dreams: Effects of task modality, length, and complexity. Frontiers in Psychology, 4, 1013. https://doi.org/10.3389/

fpsyg.2013.01013

Fasano, A. (2012). Zonulin, regulation of tight junctions, and autoimmune diseases. Annals of the New York Academy of Sciences, 1258(1), 25-33. https://doi.org/10.1111/j.1749-6632.2012.06538.x

Femi, L. G. (1978). EEG biofeedback, multichannel synchrony training, and attention. In A. Suggerman & R. E. Tarter (Eds.), Expanding dimensions of consciousness (pp. 155-182). New York: Springer Verlag.

Fenster, R. J., Lebois, L. A. M., Ressler, K. J., & Suh, J. (2018). Brain circuit dysfunction in post-traumatic stress disorder: From mouse to man. Nature Reviews. Neuroscience, 19(9), 535-551. https://doi.org/10.1038/s41583-018-0039-7

Finkelhor, D., Shattuck, A., Turner, H., & Hamby, S. (2015). A revised inventory of Adverse Childhood Experiences. Child Abuse & Neglect, 48, 13-21. https://doi.org/10.1016/j.chiabu.2015.07.011

Fiske, S. T., & Taylor, S. E. (1991). Social cognition (2nd ed.). New York: McGraw-Hill.

Forbes, D., Bisson, J. I., Monson, C. M., & Berliner, L. (2020). Effective treatments for PTSD: Practice guidelines from the International Society for Traumatic Stress Studies (3rd ed.). New York: The Guilford Press.

Frankl, V. E. (1962). Man's search for meaning: An Introduction to logotherapy. Boston: Beacon Press.

Freeman, W. J. (2005). A field-theoretic approach to understanding scale-free neocortical dynamics. Biological Cybernetics, 92(6), 350-359. https://doi.org/10.1007/s00422-005-0563-1

Freeman, W. J., Ahlfors, S. P., & Menon, V. (2009). Combining fMRI with EEG and MEG in order to relate patterns of brain activity to cognition. International journal of Psychophysiology, 73(1), 43-52. https://doi.org/10.1016/j.ijpsycho.2008.12.019

Gapen, M., van der Kolk, B. A., Hamlin, E., Hirshberg, L., Suvak, M., & Spinazzola, J. (2016). A pilot study of neurofeedback for chronic PTSD. Applied Psychophysiology and Biofeedback, 41(3), 251-261.

https://doi.org/10.1007/s10484-015-9326-5

Gilbert, M. E., & Zoeller, R. T. (2010). Thyroid hormones-impact on the developing brain: Possible mechanisms of neurotoxicity. In G. J. Harry & H. A. Tilson (Eds.), Neurotoxicology (pp. 79-111). New York, NY: Informa Healthcare.

Gold, C., Fachner, J., & Erkkilä, J. (2013). Validity and reliability of electroencephalographic frontal alpha asymmetry and frontal midline theta as biomarkers for depression. Scandinavian Journal of Psychology, 54(2), 118-126. https://doi.org/10.1111/sjop.12022

Goleman, D. (1988). The meditative mind: The varieties of meditative experience. J. R. Tarcher, Inc.

Goswami, A. (1995). The self-aware universe. Los Angeles: TarcherPerigee.

Graap, K., & Freides, D. (1998). Regarding the database for the Peniston alpha-theta EEG biofeedback protocol. Applied Psychophysiology and Biofeedback, 23(4), 265-275. https://doi.org/10.1023/a:1022265716026

Graf, W. D., Nagel, S. K., Epstein, L. G., Miller, G., Nass, R., & Larriviere, D. (2013). Pediatric neuroenhancement: Ethical, legal, social, and neurodevelopmental implications. Neurology, 80(13), 1251-1260. https://doi.org/10.1212/WNL.0b013e318289703b

Gottfredson, M. R., & Hirschi, T. (1990). A general theory of crime. Stanford, CA: Stanford University Press. http://www.sup.org/books/title/?id=2686

Green, E., & Green, A. (1977). Beyond biofeedback. San Francisco: Delacorte Press.

Green, E. E., Green, A. M., & Walters, E. D. (1974). Alpha-theta biofeedback training. Journal of Biofeedback, 2, 7-13.

Grinker, R. R., & Spiegel, J. P. (1945). Men under stress. Philadelphia, PA: Blakiston Company.

Grof, S. (1993). The holotropic mind. New York, NY: Harper Collins.

Gruzelier, J. (2009). A theory of alpha/theta neurofeedback, creative performance enhancement, long distance functional connectivity and psychological integration. Cognitive Processing, 10(Suppl 1), S101-S109. https://doi.org/10.1007/s10339-008-0248-5

Gunkelman, J., & Cripe, C. (2008) Clinical outcomes in addiction: A neurofeedback case series. Biofeedback, 36(4), 152-156.

Hanh, T. N. (2007). Living Buddha, living Christ. New York, NY: Riverhead Books.

Hantsoo, L., Jašarević, E., Criniti, S., McGeehan, B., Tanes, C., Sammel, M. D., Elovitz, M. A., Compher, C., Wu, G., & Epperson, C. N. (2019). Childhood adversity impact on gut microbiota and inflammatory response to stress during pregnancy. Brain, Behavior, and Immunity, 75, 240-250. https://doi.org/10.1016/j.bbi.2018.11.005

Harlow, H. F. (1958). The nature of love. American Psychologist, 13(12), 673-685. https://doi.org/10.1037/h0047884

Harrow, M., & Jobe, T. H. (2007). Factors involved in outcome and recovery in schizophrenia patients not on antipsychotic medications: A 15-year multifollow-up study. The Journal of Nervous and Mental Disease, 195(5), 406-414. https://doi.org/10.1097/01.nmd.0000253783.32338.6e

Healy, D. (2004). Let them eat Prozac. The unhealthy relationship between the pharmaceutical industry and depression. New York University Press.

Heller, W., Nitschke, J. B., Etienne, M. A., & Miller, G. A. (1997). Patterns of regional brain activity differentiate types of anxiety. Journal of Abnormal Psychology, 106(3), 376-385. https://doi.org/10.1037//0021-843x.106.3.376

Hirotsu, C., Tufik, S., & Andersen, M. L. (2015). Interactions between sleep, stress, and metabolism: From physiological to pathological conditions. Sleep Science (Sao Paulo, Brazil), 8(3), 143-152. https://doi.org/10.1016/j.slsci.2015.09.002

Hoffman, D. (2019). The case against reality: Why evolution hid the truth from our eyes. New York: W. W. Norton & Company.

Holmes, T. H., & Rahe, R. H. (1967). The Social Readjustment Rating Scale. Journal of Psychosomatic Research, 11(2), 213-218. https://

doi.org/10.1016/0022-3999(67)90010-4

Huecker, M. R., King, K. C., Jordan, G. A., & Smock, W. (2022). Domestic violence. In StatPearls. StatPearls Publishing.

Hyman, M. (2020). Food fix: How to save our health, our economy, our communities and our planet-one bite at a time. New York: Little, Brown Spark.

Huxley, A. (1946). The perennial philosophy (1st ed). London: Chatto and Windus.

Insel, T. R. (2015). The NIMH experimental medicine initiative. World Psychiatry (WPA), 14(2), 151-153. https://doi.org/10.1002/wps.20227

Isherwood, C., & Prabhavananda, S. (1969). How to know God: The yoga aphorisms of Patanjali. New York: Mentor.

Jeong, H. J., Durham, E. L., Moore, T. M., Dupont, R. M., McDowell, M., Cardenas-Iniguez, C., Micciche, E. T., Berman, M. G., Lahey, B. B., & Kaczkurkin, A. N. (2021). The association between latent trauma and brain structure in children. Translational Psychiatry, 11(1), 240. https://doi.org/10.1038/s41398-021-01357-z

Johnson, M., & Bodenhamer-Davis, E. (2017). The therapeutic crossover in A/T training neurofeedback: Temporal and spectral components of access to levels of consciousness. In A. Martins-Mourao & C. Kerson (Eds.), Alpha-theta neurofeedback in the 21st century: A handbook for clinicians and researchers (pp. 141-172). Murfreesboro, TN: Foundation for Neurofeedback and Neuromodulation Research.

Jokić-Begić, N., & Begić, D. (2003). Quantitative electroencephalogram (qEEG) in combat veterans with post-traumatic stress disorder (PTSD). Nordic Journal of Psychiatry, 57(5), 351-355. https://doi.org/10.1080/08039480310002688

Jung, C. (1973). Sychronicity: An acausal connecting principle. Princeton, NJ: Princeton University Press.

Kahneman, D. (2013). Thinking fast and slow. New York: Farrar, Straus and Giroux.

Kakumanu, R. J., Nair, A. K., Venugopal, R., Sasidharan, A., Ghosh, P. K., John, J. P., Mehrotra, S., Panth, R., & Kutty, B. M. (2018). Dissociating meditation proficiency and experience dependent EEG changes during traditional Vipassana meditation practice. Biological Psychology, 135, 65-75. https://doi.org/10.1016/j.biopsycho.2018.03.004

Kamiya, J. (1969). Operant control of the EEG alpha rhythms and some of its reported effect on consciousness. In C. Tart (Ed.), Altered states of consciousness (pp. 489-501). New York, NY: John Wiley & Sons.

Kang, H. J., Voleti, B., Hajszan, T., Rajkowska, G., Stockmeier, C. A., Licznerski, P., Lepack, A., Majik, M. S., Jeong, L. S., Banasr, M., Son, H., & Duman, R. S. (2012). Decreased expression of synapse-related genes and loss of synapses in major depressive disorder. Nature Medicine, 18(9), 1413-1417. https://doi.org/10.1038/nm.2886

Kasamatsu, A., & Hirai, T. (1969). An EEG study on the Zen meditation. In C. Tart (Ed.), Altered states of consciousness. New York, NY: John Wiley & Sons.

Kelly, M. J. (1997). Native Americans, neurofeedback and substance abuse theory: Three-year outcome of alpha/theta neurofeedback training in the treatment of problem drinking among Dine' (Navajo) people. Journal of Neurotherapy, 2(3), 24-60.

Kerr, C. E., Jones, S. R., Wan, Q., Pritchett, D. L., Wasserman, R. H., Wexler, A., Villanueva, J. J., Shaw, J. R., Lazar, S. W., Kaptchuk, T. J., Littenberg, R., Hämäläinen, M. S., & Moore, C. I. (2011). Effects of mindfulness meditation training on anticipatory alpha modulation in primary somatosensory cortex. Brain Research Bulletin, 85(3-4), 96-103. https://doi.org/10.1016/j.brainresbull.2011.03.026

Kerr, C. E., Sacchet, M. D., Lazar, S. W., Moore, C. I., & Jones, S. R. (2013). Mindfulness starts with the body: Somatosensory attention and top-down modulation of cortical alpha rhythms in mindfulness meditation. Frontiers in Human Neuroscience, 7, 12. https://doi.org/10.3389/fnhum.2013.00012

Kerson, C., Sherman, R. A., & Kozlowski, G. P. (2009). Alpha suppression and symmetry training for generalized anxiety symptoms. Journal of Neurotherapy, 13, 146-155. https://doi.org/10.1080/10874200903107405

Khan, A., & Brown, W. A. (2015). Antidepressants versus placebo in major depression: an overview. World Psychiatry (WPA), 14(3), 294-300. https://doi.org/10.1002/wps.20241

Kirk, I. J., & Mackay, J. C. (2003). The role of theta-range oscillations in synchronising and integrating activity in distributed mnemonic networks. Cortex, 39(4-5), 993-1008. https://doi.org/10.1016/s0010-9452(08)70874-8

Klimesch, W. (1996). Memory processes, brain oscillations and EEG synchronization. International Journal of Psychophysiology, 24(1-2), 61-100. https://doi.org/10.1016/s0167-8760(96)00057-8

Klurfeld, Z. B., Buqo, T., Sanderson, W. C., & Ward-Ciesielski, E. F. (2020). Comparing the nature of grief and growth in bereaved, divorced, and unemployed individuals. Journal of Affective Disorders, 274, 1126-1133. https://doi.org/10.1016/j.jad.2020.06.008

Knyazev, G. G. (2012). EEG delta oscillations as a correlate of basic homeostatic and motivational processes. Neuroscience and Biobehavioral Reviews, 36(1), 677-695. https://doi.org/10.1016/j.neubiorev.2011.10.002

Koch, C. (2004). The quest for consciousness: A neurobiological approach. Englewood, CO: Roberts and Company.

Koch, C. (2019). The feeling of life itself: Why consciousness is widespread but can't be computed. Cambridge, MA: The MIT Press.

Kubie, L. S. (1943). The use of induced hypnagogic reveries in the recovery of repressed amnesic data. Bulletin of the Menninger Clinic, 7, 172-182.

Kubler-Ross, E., & Kessler, D. (2005). On grief and grieving: Finding the five stages of loss. New York: Scribner's & Company.

Kuhn, T. S. (1970). The structure of scientific revolutions. Chicago: The University of Chicago Press.

LaBerge, S., & Rheingold, H. (1991). Exploring the world of lucid dreaming. New York: Ballantine Books.

Langley, C., Armand, S., Luo, Q., Savulich, G., Segerberg, T., Søndergaard, A., Pedersen, E. B., Svart, N., Overgaard-Hansen, O., Johansen, A., Borgsted, C., Cardinal, R. N., Robbins, T. W., Stenbæk, D. S., Knudsen, G. M., & Sahakian, B. J. (2023). Chronic escitalopram in healthy volunteers has specific effects on reinforcement sensitivity: A double-blind, placebo-controlled semi-randomised study. Neuropsychopharmacology, 48(4), 664-670. https://doi.org/10.1038/s41386-022-01523-x

Larsen, S., & Sherlin, L. (2013). Neurofeedback: An emerging technology for treating central nervous system dysregulation. The Psychiatric Clinics of North America, 36(1), 163-168. https://doi.org/10.1016/j.psc.2013.01.005

LeDoux, J. (1996). The emotional brain: The mysterious underpinnings of emotional life. New York: Simon & Schuster.

LeDoux, J. (2002). The synaptic self: How our brains became who we are. New York, NY: Viking.

LeDoux, J. (2015). Anxious: Using the brain to understand and treat fear and anxiety. New York, NY: Viking.

López-Tarrida, Á. D. C., de Diego-Cordero, R., & Lima-Rodríguez, J. S. (2021). Spirituality in a doctor's practice: What are the issues?. Journal of Clinical Medicine, 10(23), 5612. https://doi.org/10.3390/jcm10235612

Lubar, J. F., & Lubar, J. (1999). Neurofeedback assessment and treatment for attention deficit/hyperactivity disorders. In Evans, J. R., & Abarbanel, A. (Eds.), Introduction to quantitative EEG and neurofeedback (pp. 243-310). New York: Academic Press.

Maier, S. F., & Seligman, M. E. (2016). Learned helplessness at fifty: Insights from neuroscience. Psychological Review, 123(4), 349-367. https://doi.org/10.1037/rev0000033

Martins-Mourao, A., & Kerson, C. (2017). Alpha-theta neurofeedback in the 21st century: A handbook for clinicians and researchers. Murfreesboro, TN: Foundation for Neurofeedback and Neuromodulation Research.

Mather, M. (2007). Emotional arousal and memory binding: An object-based framework. Perspectives on Psychological Science, 2(1), 33-52. https://doi.org/10.1111/j.1745-6916.2007.00028.x

McGilchrist, I. (2009). The master and his emissary: The divided brain and the making of the western world. New Haven: Yale University Press.

Meehan, T. P., & Bressler, S. L. (2012). Neurocognitive networks: Findings, models, and theory. Neuroscience and Biobehavioral Reviews, 36(10), 2232-2247. https://doi.org/10.1016/j.neubiorev.2012.08.002

Menon, V. (2011). Large-scale brain networks and psychopathology: A unifying triple network model. Trends in Cognitive Sciences, 15(10), 483-506. https://doi.org/10.1016/j.tics.2011.08.003

Menon, V. (2015a). Large-scale functional brain organization. In A. W. Toga (Ed.), Brain mapping: An encyclopedic reference (pp. 449-459). Academic Press: Elsevier.

Menon, V. (2015b). Salience network. In A. W. Toga (Ed.), Brain mapping: An encyclopedic reference (pp. 597-611). Academic Press: Elsevier.

Miller, P. (2016). Itinerancy between attractor states in neural systems. Current Opinion in Neurobiology, 40, 14-22. https://doi.org/10.1016/j.conb.2016.05.005

Mitchell, D. J., McNaughton, N., Flanagan, D., & Kirk, I. J. (2008). Frontal-midline theta from the perspective of hippocampal "theta". Progress in Neurobiology, 86(3), 156-185. https://doi.org/10.1016/j.pneurobio.2008.09.005

Mittal, R., Debs, L. H., Patel, A. P., Nguyen, D., Patel, K., O'Connor, G., Grati, M., Mittal, J., Yan, D., Eshraghi, A. A., Deo, S. K., Daunert, S., & Liu, X. Z. (2017). Neurotransmitters: The critical modulators regulating gut-brain axis. Journal of Cellular Physiology, 232(9), 2359-2372. https://doi.org/10.1002/jcp.25518

Miura, H., Ozaki, N., Sawada, M., Isobe, K., Ohta, T., & Nagatsu, T. (2008). A link between stress and depression: Shifts in the balance between the kynurenine and serotonin pathways of tryptophan metabolism and the etiology and pathophysiology of depression. Stress (Amsterdam, Netherlands), 11(3), 198-209. https://doi.org/10.1080/10253890701754068

Mogilevski, T. (2021). The bi-directional role of the gut-brain axis in inflammatory and other gastrointestinal diseases. Current Opinion in Gastroenterology, 37(6), 572-577. https://doi.org/10.1097/MOG.0000000000000779

Montenigro, P. H., Alosco, M. L., Martin, B. M., Daneshvar, D. H., Mez, J., Chaisson, C. E., Nowinski, C. J., Au, R., McKee, A. C., Cantu, R. C., McClean, M. D., Stern, R. A., & Tripodis, Y. (2017). Cumulative head impact exposure predicts later-life depression, apathy, executive dysfunction, and cognitive impairment in former high school and college football players. Journal of Neurotrauma, 34(2), 328-340. https://doi.org/10.1089/neu.2016.4413

Moore, J. P., & Trudeau, D. L. (1998). Alpha theta brainwave biofeedback is not specific to the production of theta/alpha crossover and visualizations. Journal of Neurotherapy, 3(1), 63.

Moore, J. P., Trudeau, D. L., Thuras, P. D., Rubin, Y., Stockley, H., & Dimond, T. (2000). Comparison of alpha-theta, alpha and EMG neurofeedback in the production of alpha-theta crossover and the occurrence of visualizations. Journal of Neurotherapy, 4(1), 29-42. https://doi.org/10.1300/J184v04n01_04

Moritani, T., Smoker, W. R., Sato, Y., Numaguchi, Y., & Westesson, P. L. (2005). Diffusion-weighted imaging of acute excitotoxic brain injury. American Journal of Neuroradiology, 26(2), 216-228.

Murkar, A. L. A., & De Koninck, J. (2018). Consolidative mechanisms of emotional processing in REM sleep and PTSD. Sleep Medicine Reviews, 41, 173-184. https://doi.org/10.1016/j.smrv.2018.03.001

Naiman, R. (2017). Dreamless: The silent epidemic of REM sleep loss. Annals of the New York Academy of Sciences, 1406(1), 77-85. https://doi.org/10.1111/nyas.13447

Nash, J. (2017). Alpha/theta training as a "state access" process. In A. Martins-Mourao & C. Kerson (Eds.), Alpha-theta neurofeedback in the 21st century: A handbook for clinicians and researchers (pp. 107-121). Murfreesboro, TN: Foundation for Neurofeedback and Neuromodulation Research.

Natale, G., Limanaqi, F., Busceti, C. L., Mastroiacovo, F., Nicoletti, F., Puglisi-Allegra, S., & Fornai, F. (2021). Glymphatic system as a gateway to connect neurodegeneration from periphery to CNS. Frontiers in Neuroscience, 15, 639140. https://doi.org/10.3389/fnins.2021.639140

National Institute of Justice (1984). Incest: The last taboo. FBI Law Enforcement Bulletin, 53(2), 15-19.

Neria, Y., & Litz, B. T. (2004). Bereavement by traumatic means: The complex synergy of trauma and grief. Journal of Loss & Trauma, 9(1), 73-87. https://doi.org/10.1080/15325020490255322

Newberg, A., D'Aquili, E., & Rause, V. (2001). Why God won't go away. New York: Ballantine Books.

Nicholson, A. A., Ros, T., Densmore, M., Frewen, P. A., Neufeld, R. W. J., Théberge, J., Jetly, R., & Lanius, R. A. (2020). A randomized, controlled trial of alpha-rhythm EEG neurofeedback in posttraumatic stress disorder: A preliminary investigation showing evidence of decreased PTSD symptoms and restored default mode and salience network connectivity using fMRI. NeuroImage. Clinical, 28, 102490. https://doi.org/10.1016/j.nicl.2020.102490

Nicholson, A. A., Rabellino, D., Densmore, M., Frewen, P. A., Steryl, D., Scharnowski, F., Théberge, J., Neufeld, R. W. J., Schmahl, C., Jetly, R., & Lanius, R. A. (2022). Differential mechanisms of posterior cingulate cortex downregulation and symptom decreases in posttraumatic stress disorder and healthy individuals using real-time fMRI neurofeedback. Brain and Behavior, 12(1), e2441. https://doi.org/10.1002/brb3.2441

Niedermeyer, E., & Lopes da Silva, F. H. (2005). Electroencephalography: Basic principles, clinical applications, and related fields. New York: Lippincott Williams & Wilkins.

Noll, R. (1985). Mental imagery cultivation as a cultural phenomenon: The role of visions in shamanism. Current Anthropology, 26(4), 443-461. https://doi.org/10.1086/203305

Norris, P. (2017). How it all began. In A. Martins-Mourao & C. Kerson, (Eds.), Alpha-Theta neurofeedback in the 21st century: A handbook for clinicians and researchers (pp. 53-74). Murfreesboro, TN:

Foundation for Neurofeedback and Neuromodulation Research.

Nunez, P. I., & Srinivasan, R. (2006). Electric fields of the brain: The neurophysics of EEG (2nd ed.). New York: Oxford University Press.

Núñez-Jaramillo, L., Herrera-Solís, A., & Herrera-Morales, W. V. (2021). ADHD: Reviewing the causes and evaluating solutions. Journal of Personalized Medicine, 11(3), 166. https://doi.org/10.3390/jpm11030166

Olson, D. H., Russell, C. S., & Sprenkle, D. H. (1989). Circumplex model: Systemic assessment and treatment of families. New York: Haworth Press.

O'Mahony, S. M., Clarke, G., Borre, Y. E., Dinan, T. G., & Cryan, J. F. (2015). Serotonin, tryptophan metabolism and the brain-gut-microbiome axis. Behavioural Brain Research, 277, 32-48. https://doi.org/10.1016/j.bbr.2014.07.027

Ousdal, O. T., Andreassen, O. A., Server, A., & Jensen, J. (2014). Increased amygdala and visual cortex activity and functional connectivity towards stimulus novelty is associated with state anxiety. PloS One, 9(4), e96146. https://doi.org/10.1371/journal.pone.0096146

Öztürk, M., Yalın Sapmaz, Ş., Kandemir, H., Taneli, F., & Aydemir, Ö. (2021). The role of the kynurenine pathway and quinolinic acid in adolescent major depressive disorder. International Journal of Clinical Practice, 75(4), e13739. https://doi.org/10.1111/ijcp.13739

Pace-Schott, E. F., Gersh, T., Silvestri, R., Stickgold, R., Salzman, C., & Hobson, J. A. (2001). SSRI treatment suppresses dream recall frequency but increases subjective dream intensity in normal subjects. Journal of Sleep Research, 10(2), 129-142. https://doi.org/10.1046/j.1365-2869.2001.00249.x

Papa, A., Lancaster, N. G., & Kahler, J. (2014). Commonalities in grief responding across bereavement and non-bereavement losses. Journal of Affective Disorders, 161, 136-143. https://doi.org/10.1016/j.jad.2014.03.018

Paray, B. A., Albeshr, M. F., Jan, A. T., & Rather, I. A. (2020). Leaky gut and autoimmunity: An intricate balance in individuals health and the diseased state. International Journal of Molecular Sciences, 21(24), 9770. https://doi.org/10.3390/ijms21249770 https://doi.org/10.3390/ijms2124977

Pargament, K. I. (2011). Spiritually integrated psychotherapy: Understanding and addressing the sacred. New York: The Guilford Press.

Passini, F. T., Watson, C. G., Dehnel, L., Herder, J., & Watkins, B. (1977). Alpha wave biofeedback training therapy in alcoholics. Journal of Clinical Psychology, 33(1), 292-299. https://doi.org/10.1002/1097-4679(197701)33:1+<292::aid-jclp2270330166>3.0.co;2-l

Pelham, W. E., Altszuler, A. R., Merrill, B. M., Raiker, J. S., Macphee, F. L., Ramos, M., Gnagy, E. M., Greiner, A. R., Coles, E. K., Connor, C. M., Lonigan, C. J., Burger, L., Morrow, A. S., Zhao, X., Swanson, J. M., Waxmonsky, J. G., & Pelham, W. E. (2022). The effect of stimulant medication on the learning of academic curricula in children with ADHD: A randomized crossover study. Journal of Consulting and Clinical Psychology, 90(5), 367-380. https://doi.org/10.1037/ccp0000725

Peniston, E. G. (1998). The Peniston-Kulkosky brainwave neurofeedback therapeutic protocol: The future psychotherapy for alcoholism/PTSD/behavioral medicine. American Academy of Experts in Traumatic Stress. https://www.aaets.org/traumatic-stress-library/the-peniston-kulkosky-brainwave-neurofeedback-therapeutic-protocol-the-future-psychotherapy-for-alcoholism-ptsd-behavioral-medicine

Peniston, E. G., & Kulkosky, P. J. (1989). Alpha-theta brainwave training and beta-endorphin levels in alcoholics. Alcoholism, Clinical and Experimental Research, 13(2), 271-279. https://doi.org/10.1111/j.1530-0277.1989.tb00325.x

Peniston, E. G., & Kulkosky, P. J. (1991). Alpha-theta brainwave neurofeedback for Vietnam veterans with combat related post-traumatic stress disorder. Medical Psychotherapy, 4, 47-60.

Peniston, E. G., & Kulkosky, P. J. (1999). Neurofeedback in the treatment of addictive disorders. In J. R. Evans & A. Abarbanel (Eds.), Introduction to quantitative EEG and neurofeedback. New York:

Academic Press.

Perlmutter, D. (2015). Brain Maker: The power of gut microbes to heal and protect your brain-for life. New York: Little, Brown Spark.

Petersen, S. E., & Posner, M. I. (2012). The attention system of the human brain: 20 years after. Annual Review of Neuroscience, 35, 73–89. https://doi.org/10.1146/annurev-neuro-062111-150525

Pigott, H. E., Cannon, R., & Trullinger, M. (2021). The fallacy of sham-controlled neurofeedback trials: A reply to Thibault and colleagues (2018). Journal of Attention Disorders, 25(3), 448-457. https://doi.org/10.1177/1087054718790802

Posner, M. J., & Raichle, M. E. (1997). Images of mind. New York: W. H. Freeman.

Post, B. C., & Wade, N. G. (2009). Religion and spirituality in psychotherapy: A practice-friendly review of research. Journal of Clinical Psychology, 65(2), 131-146. https://doi.org/10.1002/jclp.20563

Prieur, J., Barbu, S., Blois-Heulin, C., & Lemasson, A. (2020). The origins of gestures and language: History, current advances and proposed theories. Biological Reviews of the Cambridge Philosophical Society, 95(3), 531-554. https://doi.org/10.1111/brv.12576

Purves-Tyson, T. D., Owens, S. J., Double, K. L., Desai, R., Handelsman, D. J., & Weickert, C. S. (2014). Testosterone induces molecular changes in dopamine signaling pathway molecules in the adolescent male rat nigrostriatal pathway. PloS One, 9(3), e91151. https://doi.org/10.1371/journal.pone.0091151

Raichle, M. E., MacLeod, A. M., Snyder, A. Z., Powers, W. J., Gusnard, D. A., & Shulman, G. L. (2001). A default mode of brain function. Proceedings of the National Academy of Sciences of the United States of America, 98(2), 676-682. https://doi.org/10.1073/pnas.98.2.676

Rasch, B., & Born, J. (2013). About sleep's role in memory. Physiological Reviews, 93(2), 681-766. https://doi.org/10.1152/physrev.00032.2012

Reichert, C. F., Deboer, T., & Landolt, H. P. (2022). Adenosine, caffeine, and sleep-wake regulation: state of the science and perspectives. Journal of Sleep Research, 31(4), e13597. https://doi.org/10.1111/jsr.13597

Robbins, J. (2008). A symphony in the brain. New York: Grove Hills.

Rogel, A., Loomis, A. M., Hamlin, E., Hodgdon, H., Spinazzola, J., & van der Kolk, B. (2020). The impact of neurofeedback training on children with developmental trauma: A randomized controlled study. Psychological trauma: Theory, Research, Practice and Policy, 12(8), 918-929. https://doi.org/10.1037/tra0000648

Rogers, C. R. (1961). On becoming a person: A therapist's view of psychotherapy. Boston New York: Houghton Mifflin Company.

Ros, T., Théberge, J., Frewen, P. A., Kluetsch, R., Densmore, M., Calhoun, V. D., & Lanius, R. A. (2013). Mind over chatter: Plastic up-regulation of the fMRI salience network directly after EEG neurofeedback. NeuroImage, 65, 324-335. https://doi.org/10.1016/j.neuroimage.2012.09.046

Rossi, E. L. (1986). The psychobiology of mind-body healing. New York: W. W. Norton.

Rostami, R., & Dehghani-Arani, F. (2015). Neurofeedback training as a new method in treatment of crystal methamphetamine dependent patients: A preliminary study. Applied Psychophysiology and Biofeedback, 40(3), 151-161. https://doi.org/10.1007/s10484-015-9281-1

Sambuco, N., Bradley, M. M., & Lang, P. J. (2022). Narrative imagery: Emotional modulation in the default mode network. Neuropsychologia, 164, 108087. https://doi.org/10.1016/j.neuropsychologia.2021.108087

Samuel, I. S., Mahmood, Z., & Saleem, S. (2014). The development of the role identification scale for adult children of alcoholic fathers. Pakistan Journal of Social and Clinical Psychology, 12(1), 3-11.

Samuelson, K. W. (2011). Post-traumatic stress disorder and declarative memory functioning: A review. Dialogues in Clinical Neuroscience, 13(3), 346-351. https://doi.org/10.31887/DCNS.2011.13.2/ksamuelson

Sapolsky, R. M. (1999). Glucocorticoids, stress, and their adverse neurological effects: Relevance to aging. Experimental Gerontology, 34(6), 721-732. https://doi.org/10.1016/s0531-5565(99)00047-9

Sapolsky, R. M. (2004). Why Zebras don't get ulcers (3rd ed.). New York: Henry Holt.

Sariaslan, A., Sharp, D. J., D'Onofrio, B. M., Larsson, H., & Fazel, S. (2016). Long-term outcomes associated with traumatic brain injury in childhood and adolescence: A nationwide swedish cohort study of a wide range of medical and social outcomes. PLoS Medicine, 13(8), e1002103. https://doi.org/10.1371/journal.pmed.1002103

Savic, I. (2015). Structural changes of the brain in relation to occupational stress. Cerebral Cortex, 25(6), 1554-1564. https://doi.org/10.1093/cercor/bht348

Saxby, E., & Peniston, E. G. (1995). Alpha-theta brainwave neurofeedback training: An effective treatment for male and female alcoholics with depressive symptoms. Journal of Clinical Psychology, 51(5), 685-693. https://doi.org/10.1002/1097-4679(199509)51:5<685::aid-jclp2270510514>3.0.co;2-k

Schnurr, P. P., Friedman, M. J., Engel, C. C., Foa, E. B., Shea, M. T., Chow, B. K., Resick, P. A., Thurston, V., Orsillo, S. M., Haug, R., Turner, C., & Bernardy, N. (2007). Cognitive behavioral therapy for posttraumatic stress disorder in women: A randomized controlled trial. JAMA, 297(8), 820-830. https://doi.org/10.1001/jama.297.8.820

Schore, A. N. (1994). Affect regulation and the origin of the self. The neurobiology of emotional development. Hillsdale, NJ: Lawrenece Erlbaum.

Schroeder, C. E., & Lakatos, P. (2009). The gamma oscillation: Master or slave?. Brain Topography, 22(1), 24-26. https://doi.org/10.1007/s10548-009-0080-y https://doi.org/10.1007/s10548-009-0080-y

Schwartz, J. M., & Begley, S. (2002). The mind and the brain: Neuroplasticity and the power of mental force. New York: Regan Books/Harper Collins Publishers.

Schwartz, M. S., & Andrasik, F. (2015). Biofeedback: A practitioner's guide (4th ed.). New York: Guilford Publications.

Scott, W. C., Kaiser, D., Othmer, S., & Sideroff, S. I. (2005). Effects of an EEG biofeedback protocol on a mixed substance abusing population. The American Journal of Drug and Alcohol Abuse, 31(3), 455-469. https://doi.org/10.1081/ada-200056807

Scott, W. (2017). Alpha/theta training applied to substance abuse and post-traumatic stress disorder. In A. Martins-Mourao & C. Kerson (Eds.), Alpha-theta neurofeedback in the 21st century: A handbook for clinicians and researchers (pp. 281-315). Murfreesboro, TN: Foundation for Neurofeedback and Neuromodulation Research.

Scott, J. G., Warber, S. L., Dieppe, P., Jones, D., & Stange, K. C. (2017). Healing journey: A qualitative analysis of the healing experiences of Americans suffering from trauma and illness. BMJ open, 7(8), e016771. https://doi.org/10.1136/bmjopen-2017-016771

Seiler, A., Fagundes, C. P., Christian, L. M. (2020). The impact of everyday stressors on the immune system and health. In A. Choukèr, (Ed.), Stress challenges and immunity in space (pp. 71-92). Springer, Cham. https://doi.org/10.1007/978-3-030-16996-1_6

Selbie, J. (2021). The physics of God: How the deepest theories of science explain religion and how the deepest truths of religion explain science. Newburyport, MA: New Page Books.

Sharma, D. (2018). Physiology of stress and its management. Journal of Medicine: Study & Research, 1, 1-5. https://doi.org/10.24966/MSR-5657/100001

Shelton, R. C., Sloan Manning, J., Barrentine, L. W., & Tipa, E. V. (2013). Assessing effects of l-methylfolate in depression management: Results of a real-world patient experience trial. The primary Care Companion for CNS Disorders, 15(4), PCC.13m01520. https://doi.org/10.4088/PCC.13m01520

Sherin, J. E., & Nemeroff, C. B. (2011). Post-traumatic stress disorder: The neurobiological impact of psychological trauma. Dialogues in Clinical Neuroscience, 13(3), 263-278. https://doi.org/10.31887/DCNS.2011.13.2/jsherin

Shonkoff, J. P., Garner, A. S., Committee on Psychosocial Aspects of Child and Family Health, Committee on Early Childhood, Adoption, and Dependent Care, & Section on Developmental and

Behavioral Pediatrics (2012). The lifelong effects of early childhood adversity and toxic stress. Pediatrics, 129(1), e232-e246. https://doi.org/10.1542/peds.2011-2663

Silverstein, D. N., & Ingvar, M. (2015). A multi-pathway hypothesis for human visual fear signaling. Frontiers in Systems Neuroscience, 9, 101. https://doi.org/10.3389/fnsys.2015.00101

Sokhadze, T. M., Cannon, R. L., & Trudeau, D. L. (2008). EEG biofeedback as a treatment for substance use disorders: Review, rating of efficacy, and recommendations for further research. Applied Psychophysiology and Biofeedback, 33(1), 1-28. https://doi.org/10.1007/s10484-007-9047-5

Solomon, B., O'Toole, L., Hong, M., & Dennis, T. A. (2014). Negative affectivity and EEG asymmetry interact to predict emotional interference on attention in early school-aged children. Brain and Cognition, 87, 173-180. https://doi.org/10.1016/j.bandc.2014.03.014

Sopp, M. R., Brueckner, A. H., Schäfer, S. K., Lass-Hennemann, J., & Michael, T. (2019). REM theta activity predicts re-experiencing symptoms after exposure to a traumatic film. Sleep Medicine, 54, 142-152. https://doi.org/10.1016/j.sleep.2018.10.030

Soutar, R. (1993). Codependency and socialization in the family of origin. [Unpublished master's thesis]. University of Arkansas.

Soutar, R. (1996). Methods in the social sciences: A critique of positivism. [Unpublished doctoral dissertation]. Oklahoma State University.

Soutar, R. (2008). Theoretical basis for the interactive self-inventory. Roswell, GA: New Mind.

Soutar, R. (2015). The automatic self: Transformation & transcendence through brainwave training. Lincoln, NE: iUniverse.

Soutar, R. (2017a). Asymmetry as a reliable measure of adult anxiety and depression. New Mind Journal, Fall, 1-4.

Soutar, R. (2017b). Perspective and method for a qEEG based two-channel bi-hemispheric compensatory model of neurofeedback training. In T. Collura, & J. Frederick (Eds.), Handbook of clinical QEEG and neurotherapy (pp. 387-403). New York, NY: Routledge.

Soutar, R. (2021). The reciprocal relationship between depression and anxiety: Archival findings 1. New Mind Journal, Fall, 1-5.

Soutar, R. (2023). When can you trust beta: Detecting electromyography contamination of the electroencephalogram signal to enhance assessment of quantitative electroencephalogram and electroencephalogram training. In D. R. Chartier, M. B. Dellinger, J. R. Evans, H. K. Budzynski (Eds.), Introduction to quantitative EEG and neurofeedback (3rd ed.) (pp. 93–102). Hoboken, NJ: Wiley & Sons. https://doi.org/10.1016/B978-0-323-89827-0.00015-2

Soutar, R., & Longo, R. (2011). Doing neurofeedback: An introduction. San Rafael: ISNR Research Foundation.

Soutar, R., & Longo, R. (2022). Doing neurofeedback: An introduction. Murfreesboro, TN: Foundation for Neurofeedback and Neuromodulation Research.

Soutar, R., Hopson, J., & Longo, R. (2017). Correlating qEEG and oxidative stress. New Mind Journal, Spring, 1-15.

Stahl, S. M. (2012). Antipsychotic polypharmacy: Never say never, but never say always. Acta Psychiatrica Scandinavica, 125(5), 349-351. https://doi.org/10.1111/j.1600-0447.2012.01841.x

Steele, K., van der Hart, O., & Nijenhuis, E. R. (2005). Phase-oriented treatment of structural dissociation in complex traumatization: Overcoming trauma-related phobias. Journal of Trauma & Dissociation (ISSD), 6(3), 11-53. https://doi.org/10.1300/J229v06n03_02

Sterman, M. B. (1996). Physiological origins and functional correlates of EEG rhythmic activities: Implications for self-regulation. Biofeedback and Self-Regulation, 21(1), 3-33. https://doi.org/10.1007/BF02214147

Sterman, M. B., & Bowersox, S. S. (1981). Sensorimotor electroencephalogram rhythmic activity: A functional gate mechanism. Sleep, 4(4), 408-422. https://doi.org/10.1093/sleep/4.4.408

Sterman, M. B., & Egner, T. (2006). Foundation and practice of neurofeedback for the treatment of epilepsy. Applied

Psychophysiology and Biofeedback, 31(1), 21-35. https://doi.org/10.1007/s10484-006-9002-x

Tagliazucchi, E. Early and contemporary human neuroimaging studies of serotonergic psychedelics. Preprints 2020, 2020050510. https://doi.org/10.20944/preprints202005.0510.v1

Taylor, S. (2018). Spiritual science. Why science needs spirituality to make sense of the world. London: Watkins Media Limited.

Teicher, M. H., Samson, J. A., Anderson, C. M., & Ohashi, K. (2016). The effects of childhood maltreatment on brain structure, function and connectivity. Nature Reviews. Neuroscience, 17(10), 652-666. https://doi.org/10.1038/nrn.2016.111

Terpou, B. A., Densmore, M., Théberge, J., Frewen, P., McKinnon, M. C., Nicholson, A. A., & Lanius, R. A. (2020). The hijacked self: Disrupted functional connectivity between the periaqueductal gray and the default mode network in posttraumatic stress disorder using dynamic causal modeling. NeuroImage. Clinical, 27, 102345. https://doi.org/10.1016/j.nicl.2020.102345

Thal, S. B., Bright, S. J., Sharbanee, J. M., Wenge, T., & Skeffington, P. M. (2021). Current perspective on the therapeutic preset for substance-assisted psychotherapy. Frontiers in Psychology, 12, 617224. https://doi.org/10.3389/fpsyg.2021.617224

Thome, J., Terpou, B. A., McKinnon, M. C., & Lanius, R. A. (2020). The neural correlates of trauma-related autobiographical memory in posttraumatic stress disorder: A meta-analysis. Depression and Anxiety, 37(4), 321-345. https://doi.org/10.1002/da.22977

Thompson, J., Stansfeld, J. L., Cooper, R. E., Morant, N., Crellin, N. E., & Moncrieff, J. (2020). Experiences of taking neuroleptic medication and impacts on symptoms, sense of self and agency: A systematic review and thematic synthesis of qualitative data. Social Psychiatry and Psychiatric Epidemiology, 55(2), 151-164. https://doi.org/10.1007/s00127-019-01819-2

Trungpa, C. (2002). Cutting through spiritual materialism. Boston: Shambala Publications.

Tsigos, C., Kyrou, I., Kassi, E., & Chrousos, G. P. (2020). Stress:

Endocrine physiology and pathophysiology. In K. R. Feingold (Eds.) et. al., Endotext. MDText.com, Inc.

Van der Kolk, B. (2014). The body keeps the score: Brain, mind, and body in the healing of trauma. New York: Viking.

van Son, D., de Rover, M., De Blasio, F. M., van der Does, W., Barry, R. J., & Putman, P. (2019). Electroencephalography theta/beta ratio covaries with mind wandering and functional connectivity in the executive control network. Annals of the New York Academy of Sciences, 1452(1), 52-64. https://doi.org/10.1111/nyas.14180

Van't Westeinde, A., & Patel, K. D. (2022). Heartfulness meditation: A yogic and neuroscientific perspective. Frontiers in Psychology, 13, 806131. https://doi.org/10.3389/fpsyg.2022.806131

Vogel, G., Foulkes, D., & Trosman, H. (1966). Ego functions and dreaming during sleep onset. Archives of General Psychiatry, 14(3), 238-248. https://doi.org/10.1001/archpsyc.1966.01730090014003

Wahbch, H., & Oken, B. S. (2013). Peak high-frequency HRV and peak alpha frequency higher in PTSD. Applied Psychophysiology and Biofeedback, 38(1), 57-69. https://doi.org/10.1007/s10484-012-9208-z

Walker, M. (2017). Why we sleep: Unlocking the power of sleep and dreams. New York: Scribner.

Wallach, M., & Kogan, N. (1965). Modes of thinking in young children. New York: Holt Rinehart & Winston.

Wang, L. A., Nguyen, D. H., & Mifflin, S. W. (2019). Corticotropin-releasing hormone projections from the paraventricular nucleus of the hypothalamus to the nucleus of the solitary tract increase blood pressure. Journal of Neurophysiology, 121(2), 602-608. https://doi.org/10.1152/jn.00623.2018

Wang, L., Gu, Y., Zhao, G., & Chen, A. (2020). Error-related negativity and error awareness in a Go/No-go task. Scientific Reports, 10(1), 4026. https://doi.org/10.1038/s41598-020-60693-0

Wang, X., & Michaelis, E. K. (2010). Selective neuronal vulnerability to oxidative stress in the brain. Frontiers in Aging Neuroscience, 2, 12. https://doi.org/10.3389/fnagi.2010.00012

Watkins, L. E., Sprang, K. R., & Rothbaum, B. O. (2018). Treating PTSD: A review of evidence-based psychotherapy interventions. Frontiers in Behavioral Neuroscience, 12, 258. https://doi.org/10.3389/fnbeh.2018.00258

Watts, A. (1961). Psychotherapy East and West. New York: Pantheon.

Wegscheider, C. S. (1989). Another chance: Hope and health for the alcoholic family. Palo Alto, CA: Science & Behavior Books.

Whitaker, R. (2010). Anatomy of an epidemic: Magic bullets, psychiatric drugs, and the astonishing rise of mental illness in America. New York: Crown Publishers.

White, N. (1999). Theories of the effectiveness of alpha-theta training for multiple disorders. In J. R. Evans & A. Abarbanel (Eds.), Introduction to quantitative EEG and neurofeedback (pp. 341-365). New York: Academic Press.

Whuttke, M. (1992). Addiction, awakening, and EEG biofeedback. Biofeedback, 20(2), 18-22.

Wilber, K. (2000). One taste: Daily reflections on integral spirituality. Boston: Shambala Publications.

Wise, A. (1997). The high-performance mind. New York: G. P. Putnam's Sons.

Wolinsky, S. (1991). Trances people live. Falls Village, CT: The Bramble Co.

Yang, C., Franciosi, S., & Brown, R. E. (2013). Adenosine inhibits the excitatory synaptic inputs to Basal forebrain cholinergic, GABAergic, and parvalbumin neurons in mice. Frontiers in Neurology, 4, 77. https://doi.org/10.3389/fneur.2013.00077

Yankouskaya, A., Denholm-Smith, T., Yi, D., Greenshaw, A. J., Cao, B., & Sui, J. (2022). Neural connectivity underlying reward and emotion-related processing: Evidence from a large-scale network analysis. Frontiers in Systems Neuroscience, 16, 833625. https://doi.org/10.3389/fnsys.2022.833625

Yuen, E. Y., Wei, J., Liu, W., Zhong, P., Li, X., & Yan, Z. (2012). Repeated stress causes cognitive impairment by suppressing

www.ingramcontent.com/pod-product-compliance
Lightning Source LLC
LaVergne TN
LVHW050909210926
848482LV00012B/384

glutamate receptor expression and function in prefrontal cortex. Neuron, 73(5), 962-977. https://doi.org/10.1016/j.neuron.2011.12.033

Zheng, A., Montez, D. F., Marek, S., Gilmore, A. W., Newbold, D. J., Laumann, T. O., Kay, B. P., Seider, N. A., Van, A. N., Hampton, J. M., Alexopoulos, D., Schlaggar, B. L., Sylvester, C. M., Greene, D. J., Shimony, J. S., Nelson, S. M., Wig, G. S., Gratton, C., McDermott, K. B., Raichle, M. E., … Dosenbach, N. U. F. (2021). Parallel hippocampal-parietal circuits for self- and goal-oriented processing. Proceedings of the National Academy of Sciences of the United States of America, 118(34), e2101743118. https://doi.org/10.1073/pnas.2101743118

Zietsch, B. P., Hansen, J. L., Hansell, N. K., Geffen, G. M., Martin, N. G., & Wright, M. J. (2007). Common and specific genetic influences on EEG power bands delta, theta, alpha, and beta. Biological Psychology, 75(2), 154-164. https://doi.org/10.1016/j.biopsycho.2007.01.004

Zisook, S., & Shear, K. (2009). Grief and bereavement: What psychiatrists need to know. World Psychiatry (WPA), 8(2), 67-74. https://doi.org/10.1002/j.2051-5545.2009.tb00217.x